Lecture Notes in Physics

Springer-Verlag
Berlin Heidelberg GmbH

The Editorial Policy for Proceedings

The series Lecture Notes in Physics reports new developments in physical research and teaching – quickly, informally, and at a high level. The proceedings to be considered for publication in this series should be limited to only a few areas of research, and these should be closely related to each other. The contributions should be of a high standard and should avoid lengthy redraftings of papers already published or about to be published elsewhere. As a whole, the proceedings should aim for a balanced presentation of the theme of the conference including a description of the techniques used and enough motivation for a broad readership. It should not be assumed that the published proceedings must reflect the conference in its entirety. (A listing or abstracts of papers presented at the meeting but not included in the proceedings could be added as an appendix.)

When applying for publication in the series Lecture Notes in Physics the volume's editor(s) should submit sufficient material to enable the series editors and their referees to make a fairly accurate evaluation (e.g. a complete list of speakers and titles of papers to be presented and abstracts). If, based on this information, the proceedings are (tentatively) accepted, the volume's editor(s), whose name(s) will appear on the title pages, should select the papers suitable for publication and have them refereed (as for a journal) when appropriate. As a rule discussions will not be accepted. The series editors and Springer-Verlag will normally not interfere with the detailed editing except in fairly obvious cases or on technical matters.

Final acceptance is expressed by the series editor in charge, in consultation with Springer-Verlag only after receiving the complete manuscript. It might help to send a copy of the authors' manuscripts in advance to the editor in charge to discuss possible revisions with him. As a general rule, the series editor will confirm his tentative acceptance if the final manuscript corresponds to the original concept discussed, if the quality of the contribution meets the requirements of the series, and if the final size of the manuscript does not greatly exceed the number of pages originally agreed upon. The manuscript should be forwarded to Springer-Verlag shortly after the meeting. In cases of extreme delay (more than six months after the conference) the series editors will check once more the timeliness of the papers. Therefore, the volume's editor(s) should establish strict deadlines, or collect the articles during the conference and have them revised on the spot. If a delay is unavoidable, one should encourage the authors to update their contributions if appropriate. The editors of proceedings are strongly advised to inform contributors about these points at an early stage.

The final manuscript should contain a table of contents and an informative introduction accessible also to readers not particularly familiar with the topic of the conference. The contributions should be in English. The volume's editor(s) should check the contributions for the correct use of language. At Springer-Verlag only the prefaces will be checked by a copy-editor for language and style. Grave linguistic or technical shortcomings may lead to the rejection of contributions by the series editors. A conference report should not exceed a total of 500 pages. Keeping the size within this bound should be achieved by a stricter selection of articles and not by imposing an upper limit to the length of the individual papers. Editors receive jointly 30 complimentary copies of their book. They are entitled to purchase further copies of their book at a reduced rate. As a rule no reprints of individual contributions can be supplied. No royalty is paid on Lecture Notes in Physics volumes. Commitment to publish is made by letter of interest rather than by signing a formal contract. Springer-Verlag secures the copyright for each volume.

The Production Process

The books are hardbound, and the publisher will select quality paper appropriate to the needs of the author(s). Publication time is about ten weeks. More than twenty years of experience guarantee authors the best possible service. To reach the goal of rapid publication at a low price the technique of photographic reproduction from a camera-ready manuscript was chosen. This process shifts the main responsibility for the technical quality considerably from the publisher to the authors. We therefore urge all authors and editors of proceedings to observe very carefully the essentials for the preparation of camera-ready manuscripts, which we will supply on request. This applies especially to the quality of figures and halftones submitted for publication. In addition, it might be useful to look at some of the volumes already published. As a special service, we offer free of charge LaTeX and TeX macro packages to format the text according to Springer-Verlag's quality requirements. We strongly recommend that you make use of this offer, since the result will be a book of considerably improved technical quality. To avoid mistakes and time-consuming correspondence during the production period the conference editors should request special instructions from the publisher well before the beginning of the conference. Manuscripts not meeting the technical standard of the series will have to be returned for improvement.

For further information please contact Springer-Verlag, Physics Editorial Department II, Tiergartenstrasse 17, D-69121 Heidelberg, Germany

Spiros Cotsakis   Gary W. Gibbons (Eds.)

Global Structure and Evolution in General Relativity

Proceedings of the First Samos Meeting
on Cosmology, Geometry and Relativity
Held at Karlovassi, Samos, Greece,
5–7 September 1994

 Springer

Editors

Spiros Cotsakis
Department of Mathematics
University of the Aegean
Karlovassi 83200, Samos, Greece

Gary W. Gibbons
DAMTP, University of Cambridge
Silver Street, Cambridge CB3 9EW, U.K.

Cataloging-in-Publication Data applied for.

Die Deutsche Bibliothek - CIP-Einheitsaufnahme

Global structure and evolution in general relativity :
proceedings of the First Samos Meeting on Cosmology,
Geometry and Relativity held at Karlovassi, Samos, Greece, 5 -
7 September 1994 / Spiros Cotsakis ; Gary W. Gibbons (ed.).

(Lecture notes in physics ; Vol. 460)
ISBN 978-3-662-14072-7   ISBN 978-3-540-49361-7 (eBook)
DOI 10.1007/978-3-540-49361-7

NE: Kōtsakēs, Spiros [Hrsg.]; Samos Meeting on Cosmology, Geometry
and Relativity <1, 1994, Neon Karlobasion>; GT

ISBN 978-3-662-14072-7

Typesetting: Camera-ready by the authors
SPIN: 10514962   55/3142-543210 - Printed on acid-free paper

Preface

This volume contains expanded versions of the invited lectures given at the First Samos Meeting on Cosmology, Geometry and Relativity. The meeting was held at Karlovassi, Samos, Greece on September 5–7, 1994 in the Department of Mathematics of the University of the Aegean. Samos will be known to many relativists as the birthplace of Pythagoras and this meeting promises to be the first in a series. The meeting was attended by about thirty participants, half from Greece and the remainder from countries as diverse as the European Union, India, South Africa and the US.

The aim of the meeting was to show the close interplay between rigorous mathematics and its use in relativity and cosmology. Rather than featuring a great variety of topics, a small number of invited speakers were given the opportunity to develop their subjects in considerable depth and detail by each being allocated two one-hour lectures.

The two themes on which the meeting was centered are reflected in this book. Part I is concerned with the application of advanced methods of analysis and partial differential equations to problems in relativity. It deals with central areas in mathematical relativity such as the initial value problem and global evolution of solutions to the Einstein equations and the Einstein vacuum constraints. Recent ideas are reviewed, for instance, the Cauchy and evolution problems for the Einstein–Yang–Mills–Boltzmann system, fluid models with finite or infinite conductivity, global evolution of a new (two–phase) model for gravitational collapse and the structure of maximal, asymptotically flat, vacuum solutions of the constraint equations which have the additional property of containing trapped surfaces. Part II focuses on geometrical–topological problems in relativity and cosmology. Two central topics in this area are addressed and anal-

Acknowledgments

As with all such endeavours, there is a long list of persons without whose help and contributed efforts the meeting would not have fulfilled its goals. We gratefully acknowledge financial support from the following sources: University of the Aegean Research Committee, Ministry of the Aegean, Ministry of Civilization, National Research and Technology Foundation and the Samos Perfecture. Peter Tselios was responsible for many of the technical aspects of the meeting.

The local member of the organizing commitee (S. Cotsakis) expresses his gratitude to Professor D. Brill for his invaluable help and encouragment during the preparation of these proceedings. His thanks also extend to Professors N. Hadjisavvas, G. Flessas and P. Leach for much help and also for patiently discussing with him many issues that arose before, during and after the meeting. He is grateful to Mr. K. Govinder for offering his expert knowledge of computing systems so willingly. They persuaded him that "life is not the complement of what it takes to organize a meeting".

We are also very grateful to the Editorial and Production Departments of Springer-Verlag for offering so willingly their expert help and opinion which resulted in the outstanding appearance of this volume.

Contents

Part I:
INITIAL VALUE AND EVOLUTION PROBLEMS

Yang-Mills Plasmas

Yvonne Choquet-Bruhat[1]

[1] Laboratoire de Gravitation et Cosmologie relativiste, Universite Paris VI, 4 Place Jussieu, Paris, France

0   Introduction

The Yang-Mills plasmas generalize to the case of a Yang-Mills type, non-abelian, charge the usual electrically charged plasmas. It has been proved recently that at very high temperature deconfined quarks and their gluon field can be effectively modeled by such a plasma (cf. [3]). Yang-Mills type plasmas, with more "elementary" particles could also have occurred in the early stages of the universe.

We consider in the first part the kinetic model of a Yang-Mills plasma: the particles have a charge which takes its values in a Lie algebra. They follow between collisions the flow lines corresponding to the average Yang-Mills field they generate through their distribution function. This function satisfies a Liouville–Vlasov equation if collisions are neglected, a Boltzmann equation if they are taken into account.

In the second part we study fluid models, with finite or infinite conductivity.

In the kinetic as well as in the fluid cases we include gravitational effects by coupling with Einstein equations.

We establish the well posedness of the obtained systems by proving local in time existence theorems for solution of the Cauchy problem with initial data on a space-like manifold. We prove global existence, under some conditions, on 4–dimensional Minkowski space time (thus excluding gravitation) for small data. Smallness means nearness to data corresponding to initial data of a solution conformal to a solution which extends to the Einstein cylinder.

For the kinetic models the conditions are that the particles have zero rest mass and the cross section satisfies some boundedness properties, verified in particular if it is zero, i.e. the model is Yang-Mills–Vlasov. Smallness means for instance smallness of f and the Yang-Mills field.

For the fluid models the condition is that the fluid is ultrarelativistic, i.e. its equation of state is $\rho = 3p$. We give an example of a solution which extends conformally to a global solution on the Einstein cylinder.

In conclusion we comment upon several open problems.

1 Kinetic Models

1.1 Definitions

1) V is a C^∞, (n+1)–dimensional manifold, g a pseudo-riemannian metric on V of signature $(- + \cdots +)$.

2) We denote by A a Yang-Mills potential on V, 1-form with values in the Lie algebra $\mathcal{G}$ of a Lie group G. In local coordinates x^α of V and a basis ε_a of $\mathcal{G}$, we have

$$A = A^a_\mu dx^\mu \varepsilon_a.$$

The Yang-Mills field (curvature of the connexion represented by A, cf. for instance [4], I) is the $\mathcal{G}$-valued 2-form

$$F = dA + \frac{1}{2}[A, A] = \frac{1}{2}F_{\lambda\mu}^a dx^\lambda \wedge dx^\mu \varepsilon_a$$

that is

$$F_{\lambda\mu}^a = \nabla_\lambda A_\mu^a - \nabla_\mu A_\lambda^a + C_{bc}^a A_\lambda^b A_\mu^c$$

where C_{bc}^a are the structure constants of $\mathcal{G}$. We suppose that $\mathcal{G}$ admits an Ad invariant scalar product, hence $c_{ba}^a = 0$.

F is of type Ad by gauge transformation and satisfies the identity

$$\hat{d}F \equiv 0 \quad \text{i.e.} \quad \hat{\nabla}_\alpha F_{\lambda\mu} + \hat{\nabla}_\mu F_{\lambda\alpha} + \hat{\nabla}_\lambda F_{\alpha\mu} \equiv 0, \tag{1.1}$$

with $\hat{\nabla} = \nabla + [A, \ ]$ the metric and gauge covariant derivative.

3) We denote by f a distribution function, in the sense of kinetic theory, on the phase space of particles with a Yang-Mills charge (an element of $\mathcal{G}$ of type Ad under gauge transformations).

For particles with arbitrary rest mass and charge, the phase space P is the fibered tensor product over V of the tangent bundle $T(V)$ with the vector bundle with base V, typical fibre $\mathcal{G}$ and group the adjoint action of G.

We will suppose (for physical reasons, also mathematically usefull) that the particles have a given rest mass, and a charge of given length in the Ad invariant, supposed positive definite, scalar product of $\mathcal{G}$, denoted by a dot. Their phase space is the subset $\mathcal{P}$ of P given by (p is the momentum and q the charge of a particle at a point x of V)

$$-g(p, p) = m^2 \geq 0, \qquad q.q = e^2 > 0 \tag{1.2}$$

If $m > 0, \mathcal{P}$ is a C^∞ submanifold of P. It is so of $\mathcal{P} - \{p = 0\}$ if $m = 0$.

The volume element $\theta = \omega_p \wedge \omega_q$ is induced by the volume element in P, the integral of $f\theta$ over an open set $U \subset \mathcal{P}$ is the average presence number of particles in U.

1.2 Vlasov–Liouville Equation

The equations of motion of a particle with Yang-Mills charge q in a Yang-Mills field F with potential A, on the manifold V with metric g, are the differential system, invariant by coordinates and gauge transformation

$$\frac{dx^\alpha}{d\sigma} = p^\alpha, \qquad \frac{dp^\alpha}{d\sigma} = P^\alpha \equiv -\Gamma^\alpha_{\lambda\mu} p^\lambda p^\mu + p^\beta q.F^\alpha_\beta$$

$$\frac{dq^a}{d\sigma} = Q^a = -c^a_{bc} p^\alpha A^b_\alpha q^c.$$

The first equations use the generalization to the Yang-Mills field of the usual Lorentz electric force, the last one called the Wong equation, expresses that the gauge covariant derivative of the charge vanishes along the trajectory.

The vector field Y defined by the right hand sides

$$Y = (p, P, Q), \quad \text{with} \quad P^\alpha = -\Gamma^\alpha_{\lambda\mu} p^\lambda p^\mu + p^\beta q.F_\beta{}^\alpha, \quad Q^a = -c^a_{bc} p^\alpha A^b_\alpha q^c$$

admits $\mathcal{P}$ as an integral submanifold: m and e are constant under the flow of Y. The volume element θ is also invariant under Y. The conservation of the number of particles moving freely (i.e. without collisions) on (V, g) in the Yang-Mills field, with distribution-function f, is expressed by the Liouville (also called Vlasov) equation

$$\mathcal{L}_Y f \equiv p^\alpha \frac{\partial f}{\partial x^\alpha} + P^\alpha \frac{\partial f}{\partial p^\alpha} + Q^a \frac{\partial f}{\partial q^a} = O.$$

We suppose $V = \mathbb{R} \times S$, we denote by x^i, $i = 1, ..., n$, local coordinates on S. To simplify computations we take for p the temporal gauge, i.e. we set:

$$g = -N^2 dt^2 + g_t \quad \text{with} \quad g_t = g_{ij}(t, x) dx^i dx^j$$

g_t a properly riemannian metric on $S_t = \{t\} \times S$.

We suppose that the particles move only into the future. We take for simplicity the gauge bundle to be trivial. The phase space $\mathcal{P}$ is then the subset of $T(V) \times \mathcal{G}$ given by

$$Np^0 = (g_{ij} p^i p^j + m^2)^{1/2}, \qquad q.q = e^2.$$

$\mathcal{P}$ is homeomorphic to $T(S) \times \mathbb{R} \times \mathcal{G}$.

In the local coordinates (x^α, p^i, q^A) on $\mathcal{P}$, with q^A local coordinates on the orbit $\mathcal{O} = \{q \epsilon \mathcal{G}, q.q = e^2\}$, the Liouville–Vlasov equation reads :

$$p^\alpha \frac{\partial f}{\partial x^\alpha} + P^i \frac{\partial f}{\partial p^i} + Q^A \frac{\partial f}{\partial q^A} = O.$$

<u>Remark.</u> If $m > 0$ we have $p^0 \geq N^{-1}$ hence x^0 is a strictly increasing parameter on a trajectory. If $m = 0$ it will be so as long as $p^0 \neq 0$, but $p^0 = 0$ is a singular point of the differential system.

1.3 Boltzmann Equation

When the particles undergo collisions the Vlasov equation must be replaced by the Boltzmann equation which contains on the right hand side an integral expressing the "balance law" of density of particles at a point with a given momentum p and charge q. If the particles undergo only binary collisions the Boltzmann equation reads

$$(\mathcal{L}_Y f)(x, p, q) = \int_{\mathcal{P}_x'} \int_{\Sigma \times \Xi} \sigma(x, p, p', q, q', p'', p''', q'', q''')$$

$$\{f(x,p'',q'')f(x,p''',q''') - f(x,p,q)f(x,p',q')\}\omega_{p'}\omega_{q'}\xi_{p''p'''}\xi_{q''q'''}$$

where the integral is taken on the direct product of $\mathcal{P}'_x$ - fiber at x of the phase space $\mathcal{P}'$ of particles with momentum p' and charge q' colliding with a particle of momentum p and charge q - with a subspace $\Sigma \times \Xi$ of $\mathcal{P}''_x \times \mathcal{P}'''_x$ fibers corresponding to the outgoing particles of momenta p'', p''' and charges q'', q'''. The subspace Σ is determined by the conservation of momentum holding for any collision

$$p + p' = p'' + p'''$$

and Ξ is determined by the charge conservation law

$$q + q' = q'' + q'''$$

while $\omega_p \omega_q$ is the coordinate and gauge invariant volume element in the fiber $\mathcal{P}_x \cong \mathbb{R}^n \times \mathcal{O}$, that is ω_q is the Ad invariant volume element in $\mathcal{O}$ and ω_p is given in the coordinates p^i by

$$\omega_p = (det g)^{1/2} \frac{dp^1 dp^2 ... dp^n}{p_0}.$$

it is smooth if $m \neq 0$, has a singularity at $p = 0$ if $m = 0$.

$\xi_{p''p'''}\xi_{q''q'''}$ is the volume element (Leray form) induced on $\Sigma \times \Xi$ by the conservation laws of momentum and charge.

In the general case Σ is diffeomorphic to a sphere in $T_x(S)$, it reduces to a point for "grazing collisions" $(p = p')$ (cf. for instance the appendix of [1]).

Taking for instance coordinates $p''^1, .\ .\ ., p''^{n-1}$ on Σ a simple calculation gives the expression of the Leray form, valid in any frame

$$\xi_{p''p'''} = | det g | \frac{dp''^1 \wedge ... \wedge dp''^{n-1}}{p''_0 p'''_n - p'''_0 p''_n}$$

If we take an appropriate orthonormal frame in $T_x(V)$ and choice of parameters on Σ, $\xi_{p''p'''}$ reduces to the canonical volume element of the unit sphere S_{n-1} multiplied by $\frac{1}{2}\frac{|p-p'|}{|p+p'|}$.

<u>Remark</u>. Since all the considered momenta have the same length the total momentum conservation implies $p^\alpha p'_\alpha = p''^\alpha p'''_\alpha$ and $\mid p - p' \mid = \mid p'' - p''' \mid$.

Analogous reasoning is used to determine Ξ and $\xi_{q''q'''}$. It obtains that Ξ is an N-2 sphere in $\mathcal{G}$ and $\xi_{q''q'''}$ the product of its canonical volume element with $\frac{1}{2}\frac{|q-q'|}{|q+q'|}$.

The function σ, called cross section, is the probability density for the considered collision. It is a function on $\mathcal{P}_x \times \mathcal{P}'_x \times \Sigma \times \Xi$. If we admit that the external fields have no influence in the collision process it is reasonable to suppose that due to Lorentz invariance, σ at a point x, depends only on the angle of p and p', that is on $P^2 = p^\alpha p_{\alpha'}$ and the "deviation angle", that is on the angle θ between $p - p'$ and and $p'' - p'''$ as well as, due to invariance in $\mathcal{G}$, analogous quantities in the q arguments.

1.4 Yang-Mills and Einstein Equations

In addition to the identities (1.1) corresponding to the first set of Maxwell equations the Yang-Mills field satisfies the Yang-Mills equations

$$\hat{\delta}F = J, \quad \text{i.e.} \quad \hat{\nabla}_\alpha F^{\alpha\beta} \equiv \nabla_\alpha F^{\alpha\beta} + [A_\alpha, F^{\alpha\beta}] = J^\beta$$

where J^β is the Yang-Mills current generated by the particles with distribution function f, i.e. :

$$J^\beta(x) = \int_{\mathcal{P}_x} p^\beta q f(x, p, q) \omega_p \omega_q$$

Lemma: If f satisfies the Boltzmann equation, the Yang-Mills current has a vanishing gauge covariant divergence:

$$\hat{\delta}J \equiv \hat{\nabla}_\beta J^\beta = 0.$$

<u>Proof</u> (cf. [2]) Various manipulations, using conservation of charge

during collisions, and the vanishing of the trace of the structure constants.

Note that a curvature satisfies the identities

$$\hat{\delta}^2 F \equiv O, \quad \text{i.e.} \quad \hat{\nabla}_\alpha \hat{\nabla}_\beta F^{\alpha\beta} \equiv O.$$

The vanishing of the covariant divergence of J makes Yang-Mills equations compatible.

The metric g satisfies the Einstein equations

$$S_{\alpha\beta} = R_{\alpha\beta} - \frac{1}{2} g_{\alpha\beta} R = T_{\alpha\beta} + \tau_{\alpha\beta} \tag{1.3}$$

where $\tau_{\alpha\beta}$ is the stress energy tensor of the Yang-Mills field and $T_{\alpha\beta}$ is generated by the distribution function

$$T_{\alpha\beta}(x) = \int_{P_x} f(x, p, q) p_\alpha p_\beta \omega_p \omega_q \tag{1.4}$$

It can be proved that if f satisfies the Boltzmann equation and A the Yang-Mills equation, then the following equation, which makes the Einstein equations compatible, holds

$$\nabla_\alpha (T^{\alpha\beta} + \tau^{\alpha\beta}) = 0. \tag{1.5}$$

1.5   Cauchy Problem. Constraint

The initial data on S for the Einstein–Yang-Mills–Boltzmann system with unknown g, A and f are: A metric $\bar{g}$ and symmetric 2 tensor K, a $\mathcal{G}$-valued 1-form a and a $\mathcal{G}$-valued vector field E on S, a function φ on $\hat{S} = T(S) \times \mathcal{O}$. A solution A, F of the system takes these initial data if, i denoting the immersion $S \to S \times \{O\} \subset V$:

$$i^* g = \bar{g}, \quad \Pi \mathcal{L}_n g = K \quad , i^* A = a, \quad F^\#.n \mid_S = E, \quad f \mid_{\hat{S}} = \varphi$$

where $F^\#$ is the contravariant $\mathcal{G}$-valued tensor associated to F, n the unit normal to S.

<u>Constraints</u> 1) The Yang-Mills equation of index zero is express-
ible in terms only of the initial data. It is therefore a $\mathcal{G}$-valued
constraint. It can be written

$$\hat{div}E = Q \equiv J.n \equiv N(0,.) \int_{\mathbf{R}^3 \times \mathcal{O}} \varphi p^0 q \omega_p \omega_q$$

$$\hat{div} \equiv div + [a,.].$$

The operator $\hat{div}$ is the L^2 adjoint of the operator $\hat{grad} \equiv grad + [a,.]$
mapping $\mathcal{G}$-valued functions on S into $\mathcal{G}$-valued 1-forms. The kernel
of $\hat{grad}$, for a given a, is the space of infinitesimal generators of 1–
parameter groups of gauge transformations leaving a invariant. In
functional spaces where the elliptic operator $\hat{div}\hat{grad}$ is surjective
when it is injective, we can construct solutions of the Yang-Mills
constraint for an arbitrary generic (i. e. admitting no 1-parameter
group of automorphisms) potential a and arbitrary Q (cf. [5]), hence
arbitrary φ, in appropriate functional spaces.

2) The Einstein constraints are the usual ones, with a given source.

1.6 Evolution Problem

We suppose that $V = S \times \mathbf{R}$. We denote $\mathcal{P}$ by $\hat{V}$, by $\hat{U}$ the subset
of $\mathcal{P}$ with projects on $U \subset V$.

1.6.1 Vlasov Equation

Suppose we are given on V a smooth and bounded as well as its
derivatives Yang-Mills potential A and a smooth metric g regularly
hyperbolic.

The corresponding Vlasov equation

$$\frac{\partial f}{\partial t} + \frac{p^i}{p^0}\frac{\partial f}{\partial x^i} + \frac{P^i}{p^0}\frac{\partial f}{\partial p^i} + \frac{Q^A}{p^0}\frac{\partial f}{\partial q^A} = 0$$

has then, <u>for $m > 0$</u>, one and only one smooth solution f on V, taking the smooth initial data φ , because, under the hypothesis made on A and g, the vector field $X = (1, \frac{p^i}{p^0}, \frac{P^i}{p^0}, \frac{Q^A}{p^0})$ is smooth and uniformly bounded as well as its first derivative, for instance in the properly riemannian metric on $\mathcal{P}$

$$ds_{\mathcal{P}}^2 = N^2 dt^2 + g_{ij}dx^i dx^j + (p^0)^{-2}g_{ij}Dp^i Dp^j + ds_{\mathcal{O}}^2,$$

where $ds_{\mathcal{O}}^2$ is thd-invariant metric on $\mathcal{O}$, and Dp^i the 1-form

$$\theta^{n+1+i} = Dp^i = dp^i + \Gamma^i_{\alpha\beta}p^\alpha dx^\beta$$

The local existence of one and only one smooth trajectory of X issued from a point of $\hat{S}_0$ is then classical. The global existence of the trajectories in the case $m > 0$ can be proved using a priori estimates of the distance of the projection in $\hat{S}$ of the points $x^I(t)$ and $x^I(0)$, and the hypothesis made on g.

From the global existence of this trajectory of X, one deduces the global existence of f with initial data φ for collision-less models. The function f has compact support in space for each time if φ has compact support.

<u>If $m = 0$</u> we can assert only local existence in time, and for an initial data φ such that

$$\{\text{Support}\varphi\} \subset \{p^0 \geq c > 0\}.$$

1.6.2   Boltzmann Equation

Though a natural functional space to solve the Boltzmann equation, and eventually prove global existence is L^1 we shall not pursue this path here because we want to couple the Boltzmann equation with the Yang-Mills–Einstein system, and L^1 is not appropriate for these equations.

We see on the expressions of the sources of the Einstein–Yang-Mills equations that f has to belong to Sobolev spaces weighted in the p directions. We can take p^0 as a weighting function on $\hat{S}_t$, since p^i/p^0 is uniformly bounded if p is regularly hyperbolic. On the other hand we do not want, at least for the local existence theorem, to make unnecessary hypothesis on the asymptotic behaviour in the space directions.

We denote by γ a properly riemannian, Sobolev regular, metric on S and we introduce the following Banach spaces for tensors on (S,γ): $H_\mu^{l,u}$ is a space of tensors of some given type whose generalized covariant derivatives of order less or equal to μ are in L^2 over every geodesic ball of a given radius δ. We set, with D the covariant derivative and ω_γ the volume element in the metric γ

$$\| v \|_{H_\mu^{l,u}} = Sup_B \| v \|_{H_\mu(B)},$$

$$\| v \|_{H_\mu(B)} = \{ \int_B \sum_{|j| \leq \mu} | D^j v |^2 \, \omega_\gamma \}^{1/2} \tag{1.6}$$

For functions on $\hat{S}$ we define a Banach space with the norm

$$\| f \|_{H_{\mu,k}^{l,u}} = Sup_{\hat{B}} \| f \|_{H_{\mu,k}(\hat{B})}$$

$$= Sup_{\hat{B}} \{ \int_{\hat{B}} \sum_{|j| \leq \mu} (p^0)^{2k+2j'+1} | D^j f |^2 \, \omega_\gamma \omega_p \omega_q \}^{1/2}$$

where j' denotes the number of derivations in the multi–index j which are taken with respect to the p variables (for collisionless models it is not necessary to include these derivations). One could prove by using a priori estimates that there exists an interval $I = [0,\ell]$ such that the Boltzmann equation has a solution $f \in C^0(I, H_{\mu,k}^{l,u})$ if A, g $\in C^0(I, H_{\mu+1}^{l,u}) \cap C^1(I, H_\mu^{l,u})$, if the cross section satisfies appropriate smoothness and boundedness conditions.

1.6.3   Yang-Mills and Einstein Equations

The evolution problem for the Yang-Mills and Einstein equations is well–posed only after choice of gauges. We choose for instance for the Yang-Mills potential the temporal gauge

$$A_0 = 0$$

and for the metric the temporal gauge indicated in paragraph 2, together with the harmonicity of the time variable, i.e. with k independent of time

$$g_{i0} = 0, \qquad N = k \mid det(g_{ij}) \mid^{1/2},$$

For a given f the Einstein–Yang-Mills system is, in these gauges, Leray hyperbolic and causal (propagation governed by the light cone). The Cauchy problem is therefore well posed if the Cauchy data, and the sources determined by f through integrals are in appropriate functional spaces. We can prove that if $f \in C^0([0, L], H^{l.u}_{\mu,k})$, with large enough μ and k the Yang-Mills–Einstein system in temporal gauges, with sources determined by f, has one and only one solution on $S \times [0, \ell]$, for sufficiently small ℓ, taking Cauchy data in $H^{l.u}_{\mu+1} \times H^{l.u}_{\mu}$.

1.6.4   Coupled System

An iteration scheme leads to a local (in time) existence theorem for the Boltzmann equation coupled with the Yang-Mills-Einstein system in temporal gauges. Usual procedures show that the obtained solution satisfies the original system if the initial data satisfy the constraints.

1.7   Global Existence on Minkowski Space–Time

The Minkovski space-time $(\mathbb{R}^4, \eta)$ is conformal to a bounded open set V of the Einstein cylinder $(S^3 \times \mathbb{R}, g)$: In canonical angular coordinates (ρ, θ, φ) on S^3 and $T \in \mathbb{R}$ where the metric g is

$$g \equiv -dT^2 + d\rho^2 + sin^2\rho(d\theta^2 + sin^2\theta d\varphi^2)$$

we have $\mathbb{R}^4 \cong V \equiv \{\rho\text{-}\pi < T < \rho+\pi\} \subset S^3 \times [-\pi,\pi]$. The diffeomorphism from $\mathbb{R}^4$ to V is given by (Penrose), when (r, θ, φ) are polar coordinates on $\mathbb{R}^3$, $t \in \mathbb{R}$,

$$T = Arctg(t + r) + Arctg(t - r)$$

$$\rho = Arctg(t + r) - Arctg(t - r)$$

We have on $V \cong \mathbb{R}^4$ that $g \equiv \Omega^2\eta$, with Ω the function

$$\Omega \equiv 2\{1 + (t + r)^2\}^{-1/2}\{1 + (t - r)^2\}^{-1/2}$$

equivalently

$$\Omega \equiv cosT + cos\rho.$$

The trace of $\Sigma_0 \equiv \mathbb{R}^3 \times \{0\}$ on V is $S_0 \equiv (S^3 - \{\rho=\pi\}) \times \{0\}$; since $\rho=\pi$ is one point of S^3 the submanifolds Σ_0 and S_0 coincide up to a set of zero measure. From Cauchy data on Σ_0 one deduces Cauchy data defined almost everywhere on the submanifold $S^3 \times \{0\}$ of the Einstein cylinder. If the system to solve is conformally invariant, or conformally regular, one deduces from a local existence theorem on the Einstein cylinder, up to the time $T = \pi$, a global existence on Minkowski space-time. The existence up to the time $T = \Pi$ can be obtained from an hypothesis of smallness of the initial data if 0 is a global solution of the system, or from an hypothesis of nearness from data corresponding to a known global solution.

1.7.1 Conformal Transformation of the Boltzmann System

Let $\underline{g} = \Omega^{-2}g$ be two conformal metrics on the manifold V. The contravariant tensors associated to a given covariant tensor on V through the metrics are related by a product with a power of Ω. We consider that a particle with 4-momentum p in (V, g) has the same covariant 4-momentum $\underline{p}$ in $(V, \underline{g})$, that is in a natural frame

$$p_\alpha = \underline{p}_\alpha, \qquad p^\alpha = \Omega^{-2}\underline{p}^\alpha$$

If the particles have zero mass, the phase space manifolds corresponding to (V, g) and $(V, \underline{g})$ are the same. We associate their distribution functions f and $\underline{f}$ by the relation

$$f(x^\alpha, p^i, q^A) = \underline{f}(x^\alpha, \underline{p}^i, q^A), \text{ with } \underline{p}^i = \Omega^2 p^i.$$

The Yang-Mills fields being identified as forms on V we deduce from the relation between the connections of two conformal metrics the following relation between the Liouville operators

$$\mathcal{L}_{\underline{Y}\underline{f}} = \Omega^2 \mathcal{L}_Y f$$

We have $\underline{N} = \Omega\, N$, $\underline{\omega}_p = \Omega^2\omega_p$; Σ and Ξ are invariant by the conformal transformation, hence

$$\mathcal{I}(\underline{f}) \equiv \int_{P'_x} \int_\Sigma \int_\Xi \{\underline{f}''\underline{f}''' - \underline{f}\,\underline{f}'\}\underline{\sigma}(x, \underline{\alpha}, \beta, \cos\theta, \cos\chi)\underline{\omega}_p\omega_q \times$$
$$\times \sin\theta d\theta d\varphi dS_{N-2} = \Omega^2 \mathcal{I}(f)$$

if we set
$$\mathcal{I}(f) = \int_{P'_x}\int_\Sigma\int_\Xi\{f''f''' - ff'\}\sigma(x, \alpha, \beta, \cos\theta, \cos\chi)\omega_{p}\omega_{q}\sin\theta d\theta d\varphi dS_{N-2}$$
with
$$\sigma(x, \alpha, \beta, \cos\theta, \cos\chi) = \underline{\sigma}(x, \Omega\alpha, \beta, \cos\theta, \cos\chi).$$

We say that the cross section $\underline{\sigma}$ is conformally regular if the right hand side of the above equation is a smooth function of Ω.

The Boltzmann equation would be conformally invariant if the cross section was $\underline{\sigma} = k \; \Theta(\cos\theta)$, i.e. independent of $\underline{\alpha}$, the total 4-momentum.

1.7.2 Conformal Transformation of the Yang-Mills Equations

It is known that in 4 dimensions the Yang-Mills operator is conformally invariant, namely

$$\hat{\underline{\nabla}}_\alpha \underline{F}^{\alpha\beta} = \Omega^4 \hat{\nabla}_\alpha F^{\alpha\beta} \tag{1.7}$$

With the identification of p_α and $\underline{p}_\alpha$ we also find

$$\underline{J}^\alpha = \Omega^4 J^\alpha \tag{1.8}$$

The Yang-Mills equation with source generated by f is therefore conformally invariant.

We deduce from these results a <u>global existence theorem on Minkowski space-time</u>, for small initial data, for the Yang-Mills–Boltzmann system, with zero rest mass particles and regular and conformally regular cross section, in particular for the Yang-Mills–Vlasov case.

1.8 Open Problems

1) <u>Global existence for small data</u>.

a) The problem is open for massive particles on Minkowski space–time, for the Yang-Mills–Vlasov model. It is often argued that at very high temperature all particles should be considered as massless, but it should be possible to extend to the Yang-Mills case the global result obtained by Glassey and Strauss for the Maxwell–Vlasov plasma.

b) Global existence on a given globally hyperbolic manifold: this problem is still open in the pure Yang-Mills case.

c) Examples of global existence in the neighbourhood of a non–zero solution.

d) Global existence with coupling to Einstein equations, result proved by Rein and Rendall in the spherically symmetric case.

2) <u>Global existence for large data</u> The problem is still open for the Maxwell–Vlasov equation on Minkowski space–time. The global existence of solutions of the pure Boltzmann equation has been proved in the non-relativistic case (without uniqueness) by Lions and Di Perna. Their result has been extended to the relativistic case.

2 Fluid Case

2.1 Definitions and Equations

The unknown on a manifold V for a gravitating Yang-Mills plasma modeled by a fluid with Yang-Mills charge are, in addition to the metric p and the potential A, the following elements:

<u>The velocity of the fluid</u>, 4-vector u tangent to V such that $u^\alpha u_\alpha = -1$.

<u>Thermodynamic quantities</u> linked with the fluid: its proper density r, its specific pressure p, enthalpy [index] i, entropy S and temperature T. They satisfy, for perfect fluids, the "thermodynamic identity"

$$TdS \equiv di - r^{-1}dp \tag{1.1}$$

and can be expressed in terms of two of them through some given "equation of state" which depends on the fluid.

The density of charge γ, function on V with values in the Lie algebra $\mathcal{G}$ and of type Ad under gauge tranformations.

The equations governing the dynamics are:

The Yang-Mills equations with source the fluid Yang-Mills current

$$\hat{\nabla}_\alpha F^{\alpha\beta} = J^\beta \tag{1.2}$$

where J is a $\mathcal{G}$-valued tangent vector to V of type Ad taken of the form, by analogy with electromagnetism,

$$J^\alpha = \gamma u^\alpha + \sigma F^{\beta\alpha} u_\beta \tag{1.3}$$

where σ is a numerical valued function, which may depend on the thermodynamic quantities, and eventually the fields. In simple models it is supposed constant, it could also depend directly on the point of space time without introducing additional mathematical difficulties.

The Einstein equations with source the sum of the Yang-Mills and fluid stress energy tensors

$$S_{\alpha\beta} = T_{\alpha\beta} + \tau_{\alpha\beta}, \tag{1.4}$$

with

$$T_{\alpha\beta} \equiv r i u_\alpha u_\beta + p g_{\alpha\beta}, \quad \text{fluid stress energy}$$

$$\tau_{\alpha\beta} \equiv -\frac{1}{4} g_{\alpha\beta} F_{\lambda\mu} . F^{\lambda\mu} + F_{\alpha\lambda} . F_\beta{}^\lambda, \quad \text{Yang} - \text{Mills stress energy}.$$

The dot denotes the Ad-invariant scalar product in $\mathcal{G}$, which we shall suppose to exist.

The proper density conservation law

$$\nabla_\alpha(r u^\alpha) = 0 \tag{1.5}$$

The <u>stress energy conservation laws</u>

$$\nabla_\alpha(T^{\alpha\beta} + \tau^{\alpha\beta}) = 0 \qquad (1.6)$$

By multiplying these equations by u^β and using proper density conservation we obtain the equation of continuity

$$-ru^\alpha\partial_\alpha i + u^\alpha\partial_\alpha p - J_\alpha.F^{\alpha\beta}u_\beta = 0.$$

and the conservation equations can then be written as <u>equations of motion</u>

$$riu^\alpha\nabla_\alpha u^\beta + (g^{\alpha\beta} + u^\alpha u^\beta)(\partial_\alpha p - J_\lambda.F^\lambda{}_\alpha) = 0 \qquad (1.7).$$

By using the thermodynamic identity the equation of continuity gives <u>entropy evolution</u>

$$rTu^\alpha\partial_\alpha S = -J_\alpha.F^{\alpha\beta}u_\beta$$

With the given expression of J the entropy equation becomes

$$rTu^\alpha\partial_\alpha S = \sigma F_\alpha{}^\lambda.F^{\alpha\beta}u_\lambda u_\beta \qquad (1.8)$$

We define as in electromagnetism the Yang-Mills electric field e by

$$e^\alpha = F^{\lambda\alpha}u_\lambda$$

it is here a $\mathcal{G}$-valued space-like tangent vector to V of type Ad. The evolution of entropy reads

$$rTu^\alpha\partial_\alpha S = \sigma e^\alpha.e_\alpha$$

therefore $u^\alpha\partial_\alpha S \geq 0$ if $\sigma \geq 0$ and the scalar product is positive. A non-positive scalar product could give for $\sigma > 0$ and some values of e an entropy decrease along the stream lines. Such scalar products may exist on Lie algebras of non–compact Lie groups.

The current must satisfy, as a consequence of the identity satisfied by the curvature F, the <u>current conservation equation</u>

$$\hat{\nabla}_\alpha J^\alpha \equiv \hat{\nabla}_\alpha(\gamma u^\alpha + \sigma e^\alpha) = 0$$

that is

$$u^\alpha \hat{\nabla}_\alpha \frac{\gamma}{r} + \frac{\sigma}{r} \hat{\nabla}_\alpha e^\alpha = 0 \tag{1.9}$$

<u>When $\sigma = 0$</u> this equation reduces to

$$u^\alpha \hat{\nabla}_\alpha(r^{-1}\gamma) = 0$$

which expresses that the ratio of the density of charge and proper density has a vanishing gauge covariant derivative along the stream lines: this of course <u>does not mean</u> that $r^{-1}\gamma$ takes a fixed value in $\mathcal{G}$ along these lines (note also that such a property would not be gauge invariant). Therefore it cannot be supposed that $r^{-1}\gamma$ is constant in the fluid, as it is generally assumed in the case of an electric charge and zero conductivity.

2.2 Vorticity and Helmholtz Equations

Generalizing the definition given by Lichnerowicz for an electrically charged fluid with a constant $\frac{\gamma}{r}$ ratio (a hypothesis which cannot be made for a non-abelian field) we define the vorticity tensor of our fluid as the 2-form

$$\pi = dC - \frac{\gamma}{r}.F, \quad \text{i.e.} \quad \Pi_{\alpha\beta} = \nabla_\alpha C_\beta - \nabla_\beta C_\alpha - \frac{\gamma}{r}.F_{\alpha\beta}, \quad \text{with } C_\alpha \equiv iu_\alpha$$

To obtain the generalization of the relativistic Helmholtz equations (cf.Y.C.B. 1958) we compute the Lie derivative with respect to the vector field C of the 2-form Π, that is (i_C denotes the interior product with C)

$$\mathcal{L}_C\Pi \equiv di_C\Pi + i_C d\Pi$$

When $\underline{\sigma = 0}$ the use of the equations of motion gives $i_C\Pi = iTdS$, hence

$$di_C\Pi = d(iT) \wedge dS$$

while, by the definition of Π,

$$d\Pi = -d(\frac{\gamma}{r}.F),$$

The use of the identity $\hat{d}F \equiv 0$ and the property of Ad-invariant scalar products

$$X.[Y,Z] = [X,Y].Z, \quad X,Y,Z \in \mathcal{G}$$

give finally the Helmholtz equations, case $\underline{\sigma = 0}$.

$$\mathcal{L}_C\Pi = d(iT) \wedge dS - i_C F. \wedge \hat{d}\frac{\gamma}{r}.$$

We see on these equations that, in the non-abelian case, in contradistinction with usual charged fluids of zero conductivity, the property of irrotationality $\Pi = 0$ is not conserved due to the non-constancy of the ratio $r^{-1}\gamma$.

2.3 Cauchy Problem Constraints and Wave Fronts

The Cauchy data on a manifold S are, in addition to the usual Einstein and Yang-Mills initial data recalled in Sect. 1, the fluid initial data, that is two scalar functions for the thermodynamic quantities, for instance $S_{|0}$ and $p_{|0}$, together with a tangent vector to S which will be the projection on S of the 4 velocity of the fluid and a $\mathcal{G}$ valued scalar function $\gamma_{|0}$. These data must satisfy the Yang-Mills and Einstein constraints.

To study the evolution one makes as usual a choice of gauges. We make the choice of temporal gauges indicated in Sect. 1 and we replace the Yang-Mills and Einstein equations by the combinations

$$\partial_0 R_{ij} - \bar{\nabla}_i R_{0j} - \bar{\nabla}_j R_{0i} = \partial_0 \rho_{ij} - \bar{\nabla}_i \rho_{0j} - \bar{\nabla}_j \rho_{0i} \qquad (2.1)$$

$$\partial_0 \hat{\nabla}_\alpha F_i^\alpha - \bar{\nabla}_i \hat{\nabla}_\alpha F_0^\alpha = \partial_0 J_i - \bar{\nabla}_i J_0 \qquad (2.2)$$

The left hand sides are respectively third order operators on A_i and g_{ij} with principal part, strictly hyperbolic if g_{ij} is properly riemannian

$$(N^{-2}\partial_{00}^2 - g^{ij}\bar{\nabla}_i\bar{\nabla}_j)\partial_0$$

The characteristic matrix of the system (2.1), (2.2), coupled with the fluid equations and current conservation, consists of a block around the diagonal corresponding to the equations (2.1), (2.2) which is a diagonal matrix with element $g^{\alpha\beta}\xi_\alpha\xi_\beta\xi_0$ and a 6×6 block around the diagonal whose determinant is, taking as independant variables S and p

$$(u^\alpha\xi_\alpha)^4 D, \qquad D \equiv (u^\alpha\xi_\alpha)^2 \frac{\partial\rho}{\partial p} - g^{\alpha\beta}\xi_\alpha\xi_\beta \qquad (2.3)$$

The cone $D = 0$ is identical with the hydrodynamic cone of uncharged perfect fluids.

In addition to these blocks the characteristic matrix contains n (the dimension of $\mathcal{G}$) lines coming from the equation (1.9).

When $\sigma = 0$ the only unknown which contributes to the principal part in the equation (1.9) is $r^{-1}\gamma$. We take it as unknown instead of γ. The contribution becomes a diagonal block on the diagonal, with element $u^\alpha\xi_\alpha$.

When $\sigma \neq 0$ the first derivatives of u and second derivatives of A appear in (1.9), they give off diagonal contribution to the characteristic matrix, though they do not change the value of the characteristic determinant.

The Leray–Ohya theory of non strict hyperbolic systems apply to our equations: the characteristic polynomial

$$P(\xi) \equiv (\xi_0 g^{\alpha\beta} \xi_\alpha \xi_\beta)^{6+3n} D(\xi)(u^\alpha \xi_\alpha)^{4+n} \tag{2.4}$$

is hyperbolic, non–strict due to appearance of multiple characteristics, if

1) $g_{0i} = 0$, g_{ij} properly riemannian, $g_{00} = -N^2 < 0$, determined as in Sect.1 , the null cone of the metric is then convex and the plane $\xi_0 = 0$ exterior to it (in the cotangent plane).

2) the vector u is time-like, and the inequality $\frac{\partial \rho}{\partial p} > 0$ is satisfied.

Remark. The vector u is given initially by its projection on S_0, the full initial 4-vector is determined by the fact that it is unitary, hence time-like. Unitarity being conserved by evolution, u remains time-like.

Under the hypothesis 1 and 2 each irreducible factor in $P(\xi)$ is indeed strictly hyperbolic and the intersection of the corresponding cones has a non empty interior. The domain of dependence is determined by the dual of this interior in the tangent plane to V, the light cone of p if $\partial \rho / \partial p \geq 1$.

One can prove by studying the cofactors of the characteristic matrix that when $\sigma = 0$ the system is equivalent to a strictly hyperbolic, quasi–diagonal, system (cf. [6] for the case of electrically charged fluids).

One deduces from these properties a local in time existence theorem for a solution of the Cauchy problem with data in usual Sobolev spaces if $\sigma = 0$, with data in Gevrey classes if $\sigma \neq 0$.

2.4 Infinite Conductivity

When the conductivity σ is large enough to be considered as infinite, or for systems of large dimensions, the electric part of the Yang-Mills field becomes negligible,

$$e_\alpha \equiv u^\beta F_{\beta\alpha} = 0 \qquad (3.1)$$

The Yang-Mills magnetic field is defined through the dual form *F as the $\mathcal{G}$ valued vector

$$h^\alpha \equiv (*F)^{\alpha\beta} u_\beta \equiv \frac{1}{2}\eta^{\alpha\beta\lambda\mu} F_{\lambda\mu} u_\beta, \qquad (3.2)$$

with η the volume 4 form. Under change of gauge h is of type Ad like F. It is orthogonal to u

$$b \equiv h^\alpha u_\alpha = 0 \qquad (3.3)$$

The equation linking the Yang-Mills potential A with h is

$$F_{\alpha\beta} \equiv \nabla_\alpha A_\beta - \nabla_\beta A_\alpha + [A_\alpha, A_\beta] = K_{\alpha\beta} \equiv \eta_{\alpha\beta\lambda\mu} h^\lambda u^\mu \qquad (3.4)$$

The Yang-Mills equation of Sect. 2 is now irrelevant, but the compatibility of the equations (3.4) imply, due to the identity $\hat{d}F \equiv 0$, the equations (for the electromagnetic case cf. [11]) called now Yang-Mills equations

$$\Phi^\beta \equiv \hat{\nabla}_\alpha (h^\alpha u^\beta - h^\beta u^\alpha) = 0 \qquad (3.5)$$

Remark. The following identity holds

$$\hat{\nabla}_\beta \Phi^\beta + \frac{1}{4}[F_{\alpha\beta}, F_{\lambda\mu} - K_{\lambda\mu}]\eta^{\lambda\mu\alpha\beta} \equiv 0 \qquad (3.6)$$

The Yang-Mills stress energy tensor reduces to

$$\tau_{\alpha\beta} \equiv \mid h \mid^2 (u_\alpha u_\beta + \frac{1}{2}g_{\alpha\beta}) - h_\alpha.h_\beta, \quad \mid h \mid^2 \equiv h^\alpha.h_\alpha.$$

The total stress energy tensor is as before the sum $\theta_{\alpha\beta}$ of this tensor and the fluid stress energy $T_{\alpha\beta}$.

The equations for a gravitating Yang-Mills plasma modeled by a fluid with infinite conductivity are, in addition to the Yang-Mills equation (3.5), the Einstein equations with source $\theta_{\alpha\beta} = T_{\alpha\beta} + \tau_{\alpha\beta}$, matter conservation, and stress energy conservation. It is straightforward to check, using the thermodynamic identity, that these equations imply entropy conservation along steam–lines

$$u_\beta \nabla_\alpha \theta^{\alpha\beta} = -r(\partial_\alpha i - r^{-1}\partial_\alpha p) = -rTu^\alpha \partial_\alpha S = 0 \qquad (3.8).$$

2.4.1 Cauchy Problem. Constraints. Wave Fronts

The Cauchy data on S are the usual Einstein initial data, a $\mathcal{G}$ valued 1 form for the induced Yang-Mills potential, two vectors tangent to S, $\bar{u}$ and $\bar{h}$ (this last one $\mathcal{G}$ valued) which will be the projection on S of the initial u and h, and two scalar functions $\bar{p}$ and $\bar{S}$. The full initial u and h are determined by the unitarity of u and the orthogonality of h with u (algebraic conditions).

In addition to the Einstein constraints the Cauchy data must satisfy the Yang-Mills constraint on S

$$\hat{\nabla}_\alpha(h^\alpha u^0 - h^0 u^\alpha) = 0$$

together with

$$F_{ij} \equiv \nabla_i A_j - \nabla_j A_i + [A_i, A_j] = \eta_{ijkl}h^l u^k$$

It can be shown (cf. [10]) that the constraints and algebraic conditions are preserved by evolution of the system written for instance in temporal gauges and with the Yang-Mills equations replaced by the following non-degenerate ones (choice proposed in [8]).

$$\hat{\nabla}_\alpha(u^\alpha h^\beta - u^\beta h^\alpha + bg^{\alpha\beta}) = 0, \qquad b = h^\lambda u_\lambda$$

<u>Remark</u>. When h is of rank greater than 1 the combination used in [9] to make magneto–hydrodynamics a well posed system does not apply because the Lorenz force is no longer orthogonal to h. Indeed

$$h_\beta \nabla_\alpha T^{\alpha\beta} = \mid h \mid^2 \hat{\nabla}_\alpha h^\alpha - h_\beta (h^\beta . \hat{\nabla}_\alpha h^\alpha) + h^\alpha (h^\beta . (\hat{\nabla}_\alpha h_\beta - \hat{\nabla}_\beta h_\alpha))$$

is not zero if h is of rank greater than 1.

2.4.2   Wave Fronts

The characteristic matrix of Einstein–Yang-Mills magneto–hydrody-namics, for instance in temporal gauge, is constituted of two blocks around the diagonal. The first one corresponding to the Einstein and Yang-Mills potential equations is diagonal with elements $g^{\alpha\beta}\xi_\alpha\xi_\beta\xi_0$ or ξ_0. The second block, corresponding to the other equations, has a determinant $D(\xi)$ which can be factorized as follows (cf. [7], [11])

$$D = (det Y)^N \Delta, \quad det Y = \xi^\alpha \xi_\alpha (u^\beta \xi_\beta)^2$$

where Δ is a 6×6 determinant. By using a proper orthonormal frame in the cotangent plane, with space axis along the principal axis of the spatial tensor $H^{ij} \equiv h^i.h^j$ one can find its explicit expression (cf. [13]). It is a third order polynomial in $(\xi_\alpha)^2$.

In the case of usual magnetohydrodynamics ($n = 1$) the tensor H is degenerate, two of its eigenvalues vanish, the polynomial Δ factorizes into an Alfen and a magneto-acoustic polynomial.

In the general case with $n > 1$ this factorization does not occur.

2.5 Global Existence for Ultrarelativistic Fluids

For very hot gases it is argued (cf [12]) that the equation of state reduces to

$$\rho = 3p \tag{4.1}$$

The stress energy tensor is then traceless:

$$Tr_g T \equiv g_{\alpha\beta}T^{\alpha\beta} = 0, \tag{4.2}$$

a property also enjoyed by the stress energy tensor of a Yang-Mills field. The equations of an ultrarelativistic Yang-Mills fluid are conformally invariant in the sense that if p and g are two conformal metrics on the 4 manifold V, $p \equiv \Omega^2 \, \underline{g}$, that is in coordinates

$$g_{\alpha\beta} = \Omega^2 \underline{g}_{\alpha\beta}$$

then the following identity holds:

$$\nabla_\alpha \theta^{\alpha\beta} \equiv \Omega^{-6} \underline{\nabla}_\alpha \theta^{\alpha\beta} \tag{4.3}$$

with

$$\underline{\theta}^{\alpha\beta} \equiv \Omega^6 \theta^{\alpha\beta} \tag{4.4}$$

The proof is by elementary computation. An analogous formula is valid in any dimension, with a different value of the exponent of Ω, for any traceless 2 tensor.

We now set, so that the vectors u and $\underline{u}$ are unitary in their respective metrics,

$$\underline{u} = \Omega \, u, \quad \text{i.e.} \quad \underline{u}^\alpha \equiv \Omega \, u^\alpha \tag{4.5}$$

and in order to satisfy (4.4)

$$\underline{\rho} = \Omega^4 \rho, \quad \underline{p} = \Omega^4 p \tag{4.6}$$

We remark that with these relations the integrals of densities ρ and $\underline{\rho}$(respectively pressures) in an open set of V and for the volume element of the corresponding metrics are equal.

For the equation of state $\rho = 3p$ we have $i = 4r^{-1}p$ and the thermodynamic identity gives

$$TdS \equiv 3r^{-1}dp - 4r^{-2}pdr$$

Therefore if r is a function of p and S

$$r^{-1}\frac{\partial r}{\partial p} = \frac{3}{4p}, \quad \text{hence} \quad r = K(S)p^{3/4}.$$

The relation (4.6) implies therefore

$$\underline{r} = \Omega^3 r$$

from which we deduce

$$\nabla_\alpha(ru^\alpha) \equiv \Omega^{-4}\underline{\nabla}_\alpha(\underline{ru}^\alpha) \tag{4.7}$$

In dimension 4 the Yang-Mills equation satisfies the identity

$$(\hat{\nabla}_\alpha F^{\alpha\beta} - J^\beta) = \Omega^{-4}(\hat{\underline{\nabla}}_\alpha \underline{F}^{\alpha\beta} - \underline{J}^\beta) \tag{4.8}$$

$$F_{\alpha\beta} = \underline{F}_{\alpha\beta}, \qquad J^\alpha = \Omega^{-4}\underline{J}^\alpha \tag{4.9}$$

The current J satisfies the relation (4.9) if

$$\gamma = \Omega^{-3}\underline{\gamma}, \qquad \sigma = \Omega^{-3}\underline{\sigma} \tag{4.10}$$

One can deduce from these conformal relations and the local existence theorem on the Einstein cylinder a global existence theorem on Minkowski space–time for "small" Cauchy data, that is remaining in a small neighbourhood of data for which the conformal solution exists on the Einstein cylinder for a time greater than π.

Remarks. 1) The case $\underline{\sigma}$ equal to a non-zero constant on Minkowski space–time is not included in our theorem, σ being then singular on the cylinder.

2) The global existence result extends without difficulty to the case of infinite conductivity.

2.5.1 Open Problem

One may hope to use the decays at infinity of sources resulting from the conformal method with the estimates obtained for vacuum Einstein equations by Christodoulou and Klainerman to extend to gravitating ultrarelativistic fluids their global existence theorem for small data.

References

[1] D. Bancel and Y. Choquet-Bruhat, *Comm. Math. Phys.* **33** (1973) 83-96.

[2] Y. Choquet-Bruhat and N. Noutchegume, *Ann I.H.P.* **55** (1991) 759-787.

and in *General Relativity and Gravitational Physics*, G. Marmo ed. World Sci.

[3] P.F. Kelly, Q. Liu, C. Lucchesi and C. Manuel, *Phys Rev D* **50** (1994) 4209–4218.

[4] Y. Choquet-Bruhat and C. DeWitt-Morette, *Analysis Manifolds and Physics*, North-Holland, Part I 1982, Part II 1989.

[5] Y. Choquet-Bruhat and D. Christodoulou, *Ann. E.N.S.* **14** (1981) 481-500.

[6] Y. Choquet-Bruhat, *Bull. Soc. Math. de France* **86** (1958) 155-175.

[7] Y. Choquet-Bruhat and T. Ruggeri, *Comm. Math. Phys.* **89** (1983) 269-275.

[8] M. van Putten, in *Waves and stability in continuous media*, ed. T. Ruggeri, to appear World Sci.

[9] A. Lichnerowicz, *Relativistic hydrodynamics and magnetohydrodynamics*, Benjamin (1967), cf. also *Waves and shock waves in magneto hydrodynamics*, Kluwer, (1994).

[10] Y. Choquet-Bruhat, *C.R. Ac. Sc. série 1* **213** (1991) 551-555 and **314** (1992) 87–91, cf. also in *Waves and stability in continuous media*, ed. T. Ruggeri, to appear World Sci.

[11] Y. Choquet-Bruhat, *Astraonautica Acta* **6** (1961) 354–365.

[12] M. Anile, *Relativistic fluids and magneto fluids*, (Cambridge University Press, Cambridge, 1989).

[13] Y. Choquet-Bruhat, *C.R.A.S. série 1* **318** (1994) 775–782.

Relativistic Fluids and Gravitational Collapse

Demetrios Christodoulou[1]

[1] Department of Mathematics, Princeton University, Princeton,
New Jersey 08544, U.S.A.

Abstract. A new model of gravitational collapse is introduced. This model has two phases, a soft phase corresponding to degenerate matter below nuclear densities and a hard phase corresponding to matter at supranuclear densities. The model contains a rich variety of features, such as interfaces across which the entropy jumps, stable static solutions corresponding to neutron stars, as well as black holes and singularities. We study the global dynamics of this model in the spherically symmetric case, in the hope of providing insight into the dynamics of real astrophysical situations.

1   General Relativity

General relativity presents us with a unified theory of space, time and gravitation, according to which spacetime is a 4-dimensional manifold M with a metric $g_{\mu\nu}$ of signature 2 whose connection represents the gravitational force. The fundamental law of the theory are the Einstein equations:

$$R_{\mu\nu} - \frac{1}{2}g_{\mu\nu}R = 8\pi T_{\mu\nu}$$

where $R_{\mu\nu}$ and R are respectively the Ricci curvature and scalar curvature of $g_{\mu\nu}$ and $T_{\mu\nu}$ is the energy-momentum tensor of matter.

In view of the twice contracted Bianchi identities

$$\nabla^\nu R_{\mu\nu} - \frac{1}{2}\partial_\mu R = 0$$

the Einstein equations imply the energy-momentum conservation laws:

$$\nabla^\nu T_{\mu\nu} = 0$$

The energy-momentum tensor is required to satisfy the positivity condition namely that, for each $x \in M$, the linear tranformation of $T_x M$:

$$v^\mu \to -T^\mu_{\ \nu} v^\nu; \quad T^\mu_{\ \nu} = g^{\mu\alpha} T_{\alpha\nu}$$

maps the closure of the open cone of future directed timelike vectors at x into itself.

Letting H_x^+ be the hyperboloid of unit future directed timelike vectors at x, the *(proper) energy density* ρ is defined by

$$\rho(x) = \inf_{u \in H_x^+} T(u, u)$$

It is a nonnegative function. The space H_x^+ being noncompact, the infimum may not be attained.

2   Perfect Fluids

A *perfect fluid* is defined to be matter whose energy tensor satisfies the following two conditions:

1. The infimum above is attained at a unique $u \in H_x^+$. The vectorfield u^μ thus defined in the support of the energy tensor is called the *fluid velocity* and we have:

$$T_{\mu\nu} u^\nu = -\rho g_{\mu\nu} u^\nu$$

The *stress* tensor $S_{\mu\nu}$ then defined by:

$$S_{\mu\nu} = T_{\mu\nu} - \rho u_\mu u_\nu; \quad u_\mu = g_{\mu\alpha} u^\alpha$$

The stress tensor at a given point $x \in M$ is a symmetric bilinear form in Σ_x, the spacelike hyperplane in $T_x M$ orthogonal to u (local simultaneous space of fluid). Since the metric induced on Σ_x is positive definite, the eigenvalues p_1, p_2, p_3 of S are defined in the usual way and are called the *principal pressures*. The positivity requirement is equivalent to the condition that each of the principal pressures is dominated in absolute value by the proper energy density:

$$|p_1|, |p_2|, |p_3| \le \rho$$

2. The principal pressures coincide:

$$p_1 = p_2 = p_3 = p$$

to a nonnegative function p, called simply *pressure*. The stress tensor is then given by:

$$S_{\mu\nu} = p(g_{\mu\nu} + u_\mu u_\nu)$$

and the energy-momentum tensor takes the form:

$$T_{\mu\nu} = (\rho + p)u_\mu u_\nu + pg_{\mu\nu}$$

3 Thermodynamics

According to the principles of thermodynamics the energy density ρ and the pressure p are functions of the two thermodynamic variables n, the *number density* of particles in the local simultaneous space of the fluid, and s, the *entropy per particle*, both taking values in the nonnegative real numbers. The differential of ρ is given by:

$$d\rho = \frac{(\rho + p)}{n}dn + n\theta ds$$

that is,

$$p = n\frac{\partial \rho}{\partial n} - \rho, \qquad \theta = \frac{1}{n}\frac{\partial \rho}{\partial s}$$

Here θ is a nonnegative function, the *(absolute) temperature*. It is assumed to vanish when and only when s vanishes. The energy-momentum conservation laws are supplemented by the particle conservation law:

$$\nabla_\mu I^\mu = 0$$

where I^μ is the particle current:

$$I^\mu = nu^\mu$$

Given ρ as a function of n and s (*equation of state*) the Einstein equations together with the particle conservation law form a complete system.

4 Barotropic Fluids

An important special case of a perfect fluid is one for which the pressure is a function of the energy density:

$$p = f(\rho)$$

Then ρ and p are functions of $\sigma = n\mu(s)$ and we have:

$$\sigma \frac{d\rho}{d\sigma} = \rho + p$$

The energy-momentum conservation laws decouple from the particle conservation law, therefore the Einstein equations form by themselves a complete system. We define:

$$V = ||V||u, \quad ||V|| = \frac{\rho + p}{\sigma}$$

Assuming that $\lim_{\sigma \to 0}(d\rho/d\sigma) = 1$, the condition that

$$\frac{dp}{d\rho} = \eta^2 \geq 0; \quad \eta : sound\ speed$$

implies: $||V|| \geq 1$. The components of the energy-momentum conservation laws along ($||$) and orthogonally ($\perp$) to u become:

$$\| : \ \nabla_\mu(G(\|V\|)V^\mu) = 0, \quad G = \frac{\sigma^2}{\rho + p}$$

$$\perp : \ \mathcal{L}_V \beta = d(\|V\|^2), \quad \beta_\mu = -g_{\mu\nu}V^\nu$$

In the case of irrotational flow: $\beta = d\phi$, where ϕ is a *time function*, $\|d\phi\| \geq 1$. The $\perp$ component is then trivially satisfied while the $\|$ component becomes the nonlinear wave equation:

$$\nabla^\mu(G(\|d\phi\|)\partial_\mu\phi) = 0$$

5 Discontinuities and Jump Conditions

Only non-spacelike hypersurfaces of discontinuity $\mathcal{B}$ can arise. Let N_μ be a covector at $x \in \mathcal{B}$ the null space of which is the tangent space of $\mathcal{B}$ at x. Then, denoting by $\Delta(\cdot)$ the jump across $\mathcal{B}$ at x, we have the jump conditions:

$$(\Delta T^{\mu\nu})N_\nu = 0, \quad (\Delta I^\mu)N_\mu = 0$$

If $T_x\mathcal{B}$ is timelike, the normal vector $N^\mu = g^{\mu\nu}N_\nu$ is spacelike and we can normalize it to be of unit length. Let N point from the side which we call *behind* and label 1 to the side which we call *ahead* and label 0. Defining $u_\perp = -g(u, N)$ the 2nd jump condition reads:

$$n_1 u_{\perp 1} = n_0 u_{\perp 0} := f$$

The quantity f is called the *particle flux*. If $f \neq 0$, the discontinuity is called a *shock*. The orientation of N is then chosen so that $f > 0$. Shocks are irreversible phenomena: the entropy per particle suffers a positive jump as we cross a shock from the side ahead to the side behind. The 1st jump condition implies the Hugoniot relation:

$$h_1^2 - h_0^2 = \left(\frac{h_1}{n_1} + \frac{h_0}{n_0}\right)(p_1 - p_0); \quad h = (\rho + p)/n \ : \ enthalpy$$

which gives the jump in entropy. In the barotropic case this reduces
to:

$$\frac{\mu_1^2}{\mu_0^2} = \frac{\sigma_1^2}{\sigma_0^2} \frac{(\rho_0 + p_0)}{(\rho_1 + p_1)} \frac{(\rho_0 + p_1)}{(\rho_1 + p_0)}$$

If $f = 0$ the discontinuity is called a *contact discontinuity*. In this
case the choice of orientation of N is merely conventional. Contact
discontinuities are reversible. If $T_x\mathcal{B}$ is null, then the normal vector
is a null vector hence belongs to $T_x\mathcal{B}$. We choose N to be future
directed. Then $u_\perp > 0$ hence $f > 0$. A hypersurface of discontinuity
which is everywhere null is called a *null shock*. There is no jump in
entropy across a null shock.

6 The Two-Phase Model

Consider the gravitational collapse of the degenerate core of a massive star. During the collapse, as long as the density of mass-energy
remains below the nuclear saturation density the equation of state
remains soft [1], that is the sound speed remains relatively low. The
adiabatic index $\gamma = (\partial \log p / \partial \log \rho)_s$ remains below the value $4/3$,
which corresponds to the ultra-relativistic lepton gas, and is the crit-
ical value for stability within the framework of the Newtonian theory
of gravitation. In fact γ decreases to a value considerably less than
$4/3$ as nuclear density is approached [2], due to the negative nuclear
pressure resulting from the diminishing surface contribution to the
energy per baryon as the nuclear size increases.

However, as soon as nuclear density is exceeded, the strong nuclear
repulsive forces come into play and the equation of state turns hard.
As the nearest neighbor separation between nucleons decreases below
twice the radius of the "hard core" the sound speed is thought to
approach the speed of light. In fact according to the results of Ref.[3],
which include three-nucleon interactions, the sound speed reaches
$0.9c$ before nuclear density is exceeded by a factor of 5. Ya. B.

Zel'dovich [4] was the first to point out that massive vector meson interactions would yield a repulsive hard core and lead to an equation of state with sound speed approaching c. These ideas where further developed by Walecka into the "relativistic mean field" [5].

In view of the above, it is not unreasonable to consider the idealized model in which the sound speed is simply zero when the density of mass-energy ρ is less than a constant ρ_0 corresponding roughly to nuclear saturation density, and equal to unity, that is to the speed of light c, when ρ exceeds ρ_0:

$$\eta^2 = \begin{cases} 0 & \text{if } \rho < \rho_0 \\ 1 & \text{if } \rho > \rho_0 \end{cases}$$

We then have a barotropic fluid:

$$p = \begin{cases} 0 & \text{if } \rho \leq \rho_0 \\ \rho - \rho_0 & \text{if } \rho > \rho_0 \end{cases}$$

with two phases, the *soft phase*, corresponding to $\rho < \rho_0$, and the *hard phase*, corresponding to $\rho > \rho_0$. We can choose the units so that $\rho_0 = 1$.

We study the dynamics of this model in the spherically symmetric case, in the hope of capturing some of the main qualitative features of the dynamics of stellar collapse. An understanding of the dynamics of our idealized model may serve as a guide to future investigations of more realistic models.

In terms of the variable σ,

$$\rho = \begin{cases} \sigma & \text{if } \sigma \leq 1 \\ \frac{1}{2}(\sigma^2 + 1) & \text{if } \sigma > 1 \end{cases}$$

$$p = \begin{cases} 0 & \text{if } \sigma \leq 1 \\ \frac{1}{2}(\sigma^2 - 1) & \text{if } \sigma > 1 \end{cases}$$

In the soft phase we have:

$$||V|| = 1, \quad G = \sigma$$

while in the hard phase:

$$||V|| = \sigma, \quad G = 1$$

In the irrotational case the equations of motion reduce in the soft phase to:

$$||d\phi|| = 1, \qquad \nabla^\mu(\sigma \partial_\mu \phi) = 0$$

and in the hard phase to:

$$||d\phi|| > 1, \qquad \nabla^2 \phi = 0$$

In the soft phase ϕ is the temporal distance function from a spacelike hypersurface and

$$T_{\mu\nu} = \rho \partial_\mu \phi \partial_\nu \phi$$

In the hard phase

$$T_{\mu\nu} = \partial_\mu \phi \partial_\nu \phi + \frac{1}{2}(||d\phi||^2 - 1)g_{\mu\nu}$$

differs from that of a massless scalar field by a "cosmological" term corresponding to a cosmological constant: $\Lambda = 4\pi\rho_0$.

The soft phase is commonly called "dust". The first study of relativistic gravitational collapse was the study of the collapse of a homogeneous dust ball by Oppenheimer and Snyder [6]. Their paper remained for a long time the only analytic study of relativistic gravitational collapse. Despite the fact that pressure is entirely neglected in this model, it was thought to give a faithful description of the formation of a black hole when the total mass of the ball is very large, for in that case the maximal density reached at the point where the black hole forms is correspondingly low.

The more general case of an inhomogeneous dust ball was analyzed in Ref.[7], where it was found that, on the contrary, no matter how large the total mass of the ball is, as long as the collapse proceeds from an initial state of low compactness, the density becomes infinite at the center of the ball before the black hole has a chance to form, thus invalidating the neglect of the pressure and casting doubt on the predictions of the model from this point on, in particular on the prediction that a black hole shall form.

7   The Scalar Field Model

If the "cosmological" term is neglected, the hard phase reduces in the irrotational case to a massless scalar field. A hydrodynamic interpretation is possible only if the gradient of the wave function is an everywhere timelike future directed vectorfield. However, even in regions where this does not hold, the scalar field continues to make sense physically as a wave field.

The scalar field model with the wave field interpretation was analyzed, in the spherically symmetric case, in Ref.[8], where it was shown that when the final Bondi mass is different from zero, as the retarded time measured by faraway observers tends to infinity, a black hole forms of mass equal to the final Bondi mass surrounded by vacuum. The rate of growth of the redshift was determined and the wave behaviour was analyzed.

In Ref.[9], a theorem was established to the effect that if on an initial future cone with vertex at the center of symmetry, there is a region bounded by two spheres such that the ratio that the mass contained in the region bears to the largest radius is large in comparison to the ratio of the radii minus 1, then a trapped region, that is a spacetime region where the future cones with vertices at the center of symmetry have negative expansion, forms in the future ending at a singular boundary. Moreover, the final Bondi mass is bounded from below by a positive number depending only on the two initial radii. The structure of the apparent horizon, that is, the boundary of the trapped region and that of the singular boundary was determined.

In Ref.[10], solutions of bounded variation were considered and a sharp criterion on the initial data was found for the avoidance of collapse, as well as another criterion for the development of singularities, complementing the results just mentioned. Moreover, the behaviour of the solutions at the center of symmetry, where the discontinuities focus, as well as the behaviour near the singular boundary, was analyzed in detail.

In Ref.[11], a family of examples was constructed, which are solutions with regular, asymtotically flat initial data, which develop naked singularities.

In Ref.[12], a theorem is demonstrated to the effect that in the space of initial conditions the subset leading to the formation of naked singularities has the following property: through each point of the subset there passes a plane which intersects the subset only at the given point, while the planes corresponding to distinct points of the subset do not intersect. Thus, the formation of naked singularities is an unstable phenomenon in the context of the spherical scalar field model.

The above work by no means exhausts the scalar field model. The fascinating discoveries of Choptuik [13], in his numerical study of the scalar field model, remain to be understood theoretically.

8   Phase Transitions

In the hydrodynamic context, each of the two phases is by itself incomplete. For, consider the irrotational case. Suppose that the initial conditions correspond to the soft phase. The function ϕ is in the soft phase the temporal distance function from a spacelike hypersurface Σ. For each point q of Σ at which Σ has negative mean extrinsic curvature there is a focal point q_* of the timelike geodesic congruence orthogonal to Σ along the geodesic from q toward the future. The mass-energy density ρ blows up at q_*, unless ρ vanishes at q. However, before ρ can become infinite it must exceed unity, therefore there must be a phase transition to the hard phase.

Suppose next that the initial conditions correspond to the hard phase. The gradient of the function ϕ is timelike future directed and of norm exceeding unity along the initial Cauchy hypersurface Σ and ϕ is a solution of the wave equation in the future of Σ. Now the gradient of ϕ may not remain timelike in the future of Σ; but before the gradient of ϕ becomes non-timelike, its norm must become less than unity, therefore there a phase transition to the soft phase must occur. Thus we may say that the soft phase turns in contraction into the hard phase and, while the hard phase turns upon expansion into the soft phase. Only the two phases taken together constitute a complete model.

The requirement that a discontinuity arise naturaly from smooth initial conditions implies that the hypersurface of discontinuity $\mathcal{B}$ is superpersonic relative to the state ahead and subsonic relative to the state behind. For our two phase model this implies that the state behind cannot belong to the soft phase. Therefore only contact discontinuities are possible within the soft phase. Also, the only type of discontinuity that may develop within the hard phase is a null shock. These null shocks are the discontinuities studied in the paper on bounded variation solutions [10]. The remaining possibilities are a timelike shock with the state ahead in the soft phase and the state behind in the hard phase, or a contact discontinuity, as part of the boundary between the two phases.

To summarize, the phase boundary has both spacelike as well as timelike components. Across a spacelike component the thermodynamic variables and the fluid velocity are continuous, the final values of one phase providing the initial data for the next phase. However, across a timelike component the fluid velocity and the thermodynamic variables suffer discontinuities, determined by the integral form of the conservation laws, that is, by the jump conditions. These discontinuities are of an irreversible character, each point of a timelike component which is crossed by a flow line being a point of increase of the entropy.

9   Spherical Symmetry

In the case of spherical symmetry the group $SO(3)$ acts as an isometry group on the spacetime manifold M. The quotient of M by the group is a 2-dimensional manifold Q with boundary. The boundary of Q corresponds to the set of fixed points of the group action which is a timelike geodesic Γ (center of symmetry). The quotient spacetime Q is endowed with a metric g_{ab} of signature 0. The metric of M takes the form:

$$g_{ab}(x)dx^a dx^b + r^2(x) \overset{\circ}{\gamma}_{AB}(y)dy^A dy^B$$

Here $\overset{\circ}{\gamma}_{AB}(y)dy^A dy^B$ is the metric of the standard unit 2-sphere expressed in arbitrary local coordinates and $r(x) = \sqrt{\frac{area(x)}{4\pi}}$ is the *area radius* of the sphere corresponding to x. The energy tensor takes the form:

$$T_{ab}(x)dx^a dx^b + r^2(x)S(x) \overset{\circ}{\gamma}_{AB}(y)dy^A dy^B$$

The Einstein equations become:

$$\nabla_a \nabla_b r = (1/2r)(1 - \partial^c r \partial_c r)g_{ab} - 4\pi r(T_{ab} - g_{ab}trT)$$

$$\nabla_b(r^2 T^{ab}) = 2r\partial^a rS$$

These imply the following equation for the Gauss curvature of Q:

$$K = (1/r^2)(1 - \partial^a r \partial_a r) + 4\pi(trT - 2S)$$

The *mass function* is defined by:

$$1 - \frac{2m}{r} = g^{ab}\partial_a r \partial_b r$$

and satisfies the equations:

$$\partial_a m = 4\pi r^2 (T_{ab} - g_{ab}trT)\partial^b r$$

For a perfect fluid,

$$T_{ab} = (\rho + p)u_a u_b + pg_{ab} \quad S = p$$

$u = u^a \partial/\partial x^a$, the flow is irrotational: $\beta = d\phi$, and, in the barotropic case,

$$\nabla^a(r^2 G(||d\phi||)\partial_a\phi) = 0$$

10 The Phase Transition from Soft to Hard

Consider initial data crresponding to the soft phase. In comoving coordinates (τ, χ) the metric of Q in the soft phase takes the form: $-d\tau^2 + e^{2\omega}d\chi^2$ and we have: $\phi = \tau$. Let $\mathcal{U}_*$ be the maximal developement of the soft phase solution. The soft phase is limited by the condition $\rho < 1$. Therefore the set

$$\mathcal{K} = \{(\tau, \chi) \in \mathcal{U}_* : \rho(\tau, \chi) \geq 1\}$$

must be excised from $\mathcal{U}_*$. But this is not all; for, causality demands that for each excluded point the domain of its influence, i.e. its causal future, be also excluded. Thus we must excise from $\mathcal{U}_*$ the set $\mathcal{J}^+(\mathcal{K})$, the causal future of $\mathcal{K}$ in $\mathcal{U}_*$. The boundary of $\mathcal{J}^+(\mathcal{K})$, consists of smooth spacelike segments Σ_i along which $\rho = 1$ and which form the spacelike part of the phase boundary, joined by pairs C_i^+, C_{i+1}^- of outgoing and incoming null segments.

The end points N_i^+ and N_i^- of the spacelike segment Σ_i at which Σ_i turns null, and which are, at the same time, the past end points of the null segments C_i^+ and C_{i+1}^- respectively, we call *boundary null points*. These are analogous to the *branch points* of minimal surface theory.

The data induced by the soft phase along the spacelike segments Σ_i, provide the initial conditions for a subsequent hard phase, determined in the future domain of dependence of Σ_i. The extension of the solution to the causal future of the boundary null points requires the solution of a *free boundary problem*. Let us confine attention to

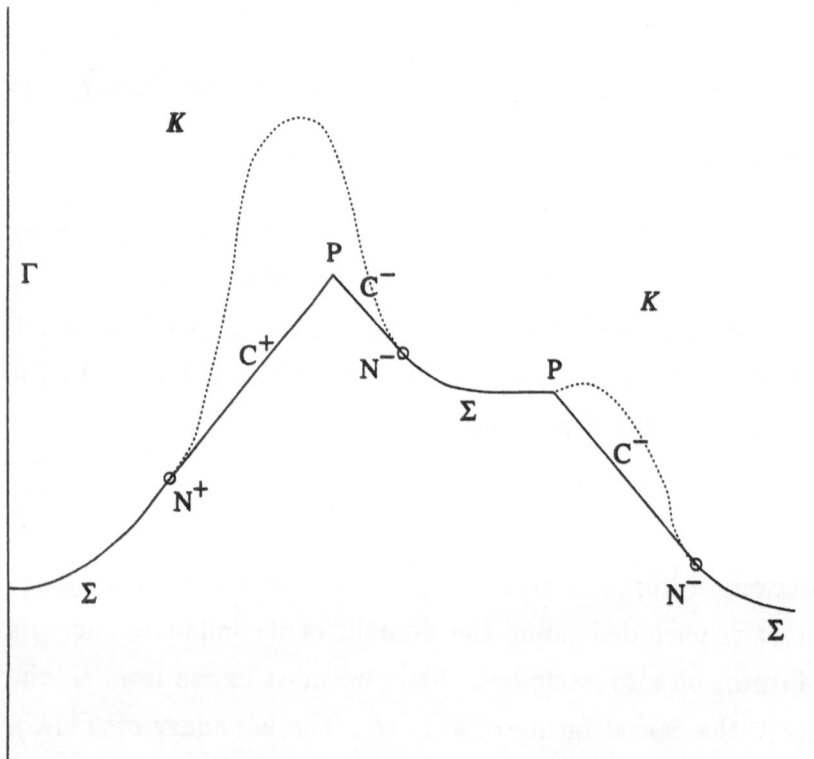

Fig. 1. The phase transition from soft to hard.

a particular outgoing boundary null point N^+. Along C^+, the outgoing null curve issuing from N^+, we have data induced from the soft phase solution, while along C^{*+}, the incoming null curve issuing from N^+, we have characteristic initial data induced from the hard phase solution in $\mathcal{D}^+(\Sigma)$. Let C^+ be given by: $\tau = \tau_+(\chi)$ and let $\partial\mathcal{K}$ be given by: $\tau = \hat{\tau}(\chi)$.

11 The Problem of Formation of a Free Phase Boundary

Find a timelike curve $\mathcal{B}$ in the soft phase domain $\mathcal{U}_0 = \{(\tau, \chi) : \tau_+(\chi) \leq \tau \leq \hat{\tau}(\chi), \chi \geq \chi_0\}$ issuing from $N^+ = (\tau_0, \chi_0)$, and a hard phase solution in $\mathcal{U}_1 = \{(u, v) : s_1 \geq u \geq v \geq 0\}$, where u, v are canonical null coordinates, i.e. coordinates whose level curves are, respectively, outgoing and incoming null curves and $u = v = s$: arc length along $\mathcal{B}$, such that identifying $\mathcal{B}$ with $\{(u, v) \in \mathcal{U}_1 : u = v\}$ by identifying points corresponding to the same value of the arc length s, we have a developement $\mathcal{U} = \mathcal{U}_1 \bigcup \mathcal{U}_0$ such that r is continuous across $\mathcal{B}$, $m_1(N^+) = m_0(N^+)$ and the jump conditions are fulfiled across $\mathcal{B}$. (It then folows that dr, m are continuous across $\mathcal{B}$ and g_{ab} is C^1 in the canonical null coordinate system).

The ratio:

$$\gamma = \frac{[(\partial r/\partial u)/(\partial\phi/\partial u)]_1}{[(\partial r/\partial u)/(\partial\phi/\partial u)]_0} \quad 0 < \gamma \leq 1$$

plays a fundamental role in the problem. The velocity of $\mathcal{B}$ relative to the soft phase flow lines is given by:

$$\beta = \frac{1 - \gamma^2}{1 + \gamma^2 - 2\gamma^2\rho_*} \quad 0 \leq \beta \leq 1$$

where ρ_* is the soft phase mass-energy density along $\mathcal{B}$. Note that $\hat{\tau}(\chi_0) = \tau_+(\chi_0) = \chi_0, \gamma(\tau_0) = \rho_*(\tau_0) = 1$. Denoting by ϕ_* the hard phase wave function along $\mathcal{B}$ we have, in general, $d\phi_*/d\tau \leq 1$ with equality at a point if and only if either $\rho_* = 1$ or $\beta = 0$ at that point.

Theorem: Let (r, ω, ρ) be a soft phase solution corresponding to smooth initial data and let $\mathcal{U}_0$ be the soft phase domain bounded by C^+, the outgoing null curve issuing from $N^+ = (0,0)$, and $\partial\mathcal{K}$, where, for the given soft phase initial data, $\rho(\tau, \chi)$ along each flow line first becomes equal to 1. Let also R be a function defined on an interval $[0, t_1]$ and representing r as a function of ϕ along C^{*+}, the incoming null curve issuing from N^+. Then there is a $\tau_1 > 0$ and a solution to the problem of formation of a free phase boundary such that $\mathcal{B}$ is a C^1 curv flow lines, is strictly timelike (i.e. $\beta < 1$) and contained in the interior of $\mathcal{U}_0$ except for its origin N^+, becoming null outgoing at N^+ ($\beta(0) = 1$). Also, on $\mathcal{U}_1$, $s_1 = s(\tau_1)$, r, m, ϕ are C^1 functions defining a genuine hard phase solution, i.e. $\rho > 1$ in $\mathcal{U}_1$ except at N^+ where $\rho = 1$.

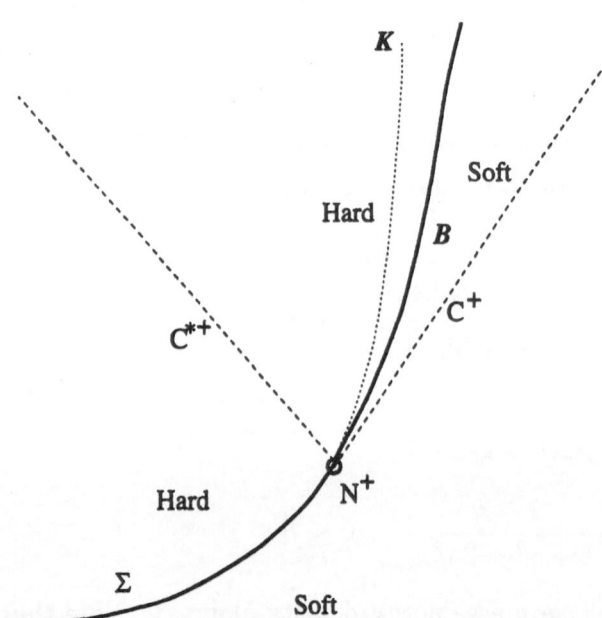

Fig. 2. Formation of a free phase boundary.

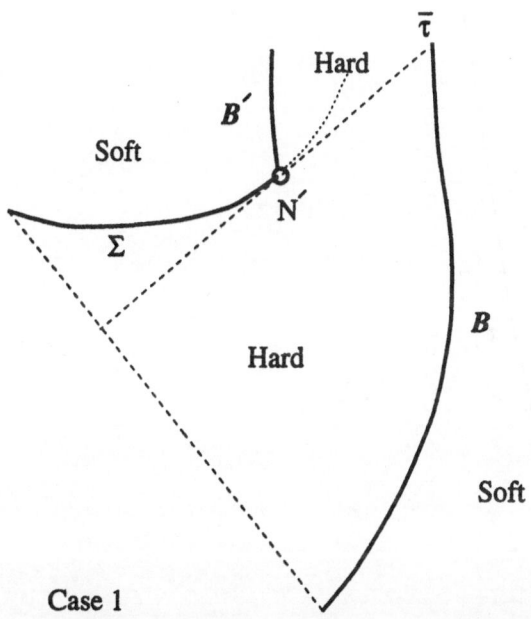

Case 1

Fig. 4. The termination of a free phase boundary. Case 1.

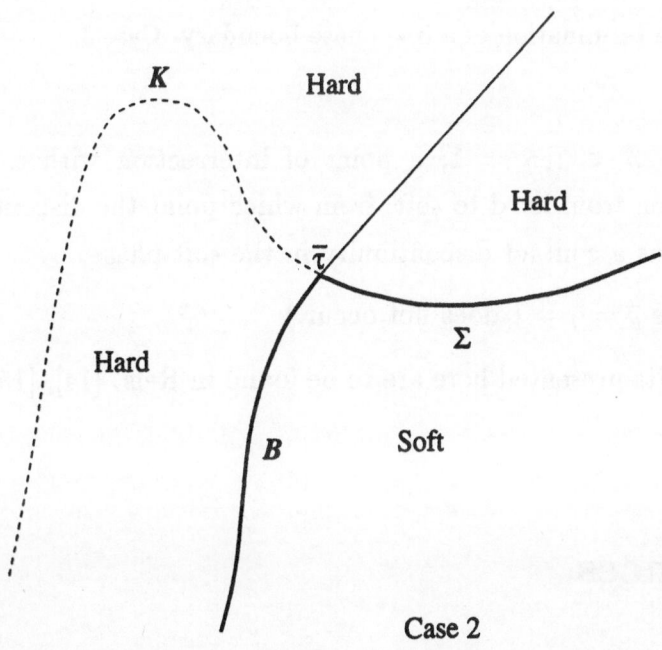

Case 2

Fig. 5. The termination of a free phase boundary. Case 2.

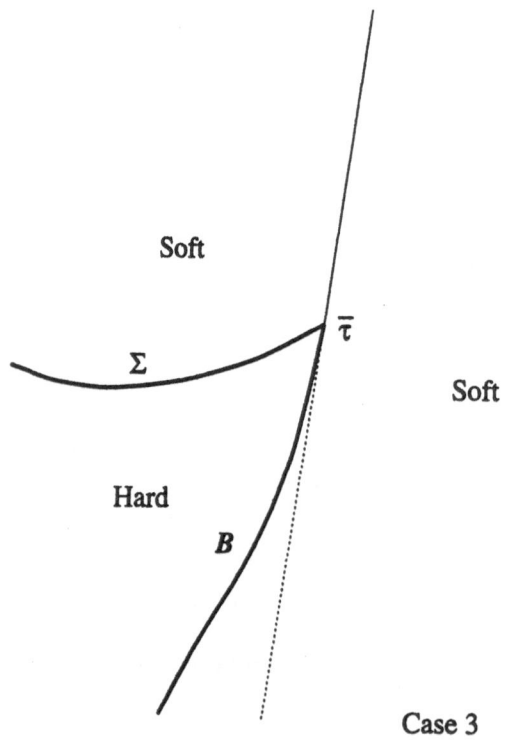

Fig. 6. The termination of a free phase boundary. Case 3.

Case 3: $\bar{\rho} < 1, \bar{\gamma} = 1$; a point of intersection with a space-like transition from hard to soft, from which point the discontinuity propagates as a contact discontinuity in the soft phase.

(The case $\bar{\rho} = \bar{\gamma} = 1$ does not occur.)

The results presented here are to be found in Refs. [14], [15], [16], and [17].

References

[1] H. A. Bethe, G. E. Brown, J. Applegate and J. M. Lattimer, "Equation of state in the gravitational collapse of stars", *Nuc. Phys.* **A324** (1979) 487.

[2] H. A. Bethe, G. E. Brown, J. Cooperstein and J. R. Wilson, "A simplified equation of state near nuclear density", *Nuc. Phys.* **A403** (1983) 625.

[3] B. Friedman and V. R. Pandharipande, "Hot and cold, nuclear and neutron matter", *Nuc. Phys.* **A361** (1981) 502.

[4] Ya. B. Zel'dovich, "The equation of state at ultrahigh densities and its relativistic limitations", *J. Exp. Theor. Phys. (U.S.S.R.)* **41** (1961) 1609 (English trans. in *Sov. Phys. JETP* **14** (1962) 1143).

[5] J. D. Walecka, "A theory of highly condensed matter", *Ann. Phys.* **83** (1974) 491.

[6] J. R. Oppenheimer and H. Snyder, "On continued gravitational contraction", *Phys. Rev.* **56** (1939) 455.

[7] D. Christodoulou, "Violation of cosmic censorship in the gravitational collapse of a dust cloud", *Commun. Math. Phys.* **93** (1984) 171.

[8] D. Christodoulou, "A mathematical theory of gravitational collapse", *Commun. Math. Phys.* **109** (1987) 613.

[9] D. Christodoulou, "The formation of black holes and singularities in spherically symmetric gravitational collapse", *Commun. Pure & Appl. Math.* **XLIV** (1991) 339.

[10] D. Christodoulou, "Bounded variation solutions of the spherically symmetric Einstein-scalar field equations", *Commun. Pure & Appl. Math.* **XLVI** (1993) 1131.

[11] D. Christodoulou, "Examples of naked singularity formation in the gravitational collapse of a scalar field", *Ann. Math.* **140** (1994) 607.

[12] D. Christodoulou, "The instability of naked singularities in the gravitational collapse of a scalar field", *Ann. Math.* (to appear).

[13] M. W. Choptuik, "Universality and scaling in gravitational collapse of a massless scalar field", *Phys. Rev. Lett.* **70** (1993) 9.

[14] D. Christodoulou, "Self-gravitating relativistic fluids: a two-phase model", *Arch. Rat. Mech. Anal.* (to appear).

[15] D. Christodoulou, "Self-gravitating relativistic fluids: The Continuation and Termination of A Free Phase Boundary", *Arch. Rat. Mech. Anal.* (to appear).

[16] D. Christodoulou, "Self-gravitating relativistic fluids: the Formation of a Free Phase boundary in the Phase transition from Soft to Hard", *Arch. Rat. Mech. Anal.* (to appear).

[17] D. Christodoulou, "Self-gravitating relativistic fluids: the Formation of a Free Phase boundary in the Phase transition from Hard to Soft", *Arch. Rat. Mech. Anal.* (to appear).

The Einstein Vacuum Constraints and Trapped Surfaces

R. Beig[1,2]

[1] Institut für Theoretische Physik, Universität Wien, Boltzmanngasse 5,
A–1090 Wien, Austria
Fax: ++43-1-317-22-20, E-mail: BEIG@PAP.UNIVIE.AC.AT

Abstract. We study maximal, asymptotically flat solutions of the constraint equations of General Relativity in vacuo. Our interest is in solutions having the additional property of containing trapped surfaces, so that the Penrose singularity theorem applies to spacetimes which Cauchy evolve from these data. We outline results, based on joint work with N. Ó Murchadha, which prove the existence of a large class of such data.

1 Introduction

We know from basic physics that the electromagnetic field of an electric charge necessarily has a long–range ("Coulombic") component. On the other hand, electromagnetism is governed by equations obeying relativistic causality. Clearly there is a tension between the notion of Coulomb fields following charges slavishly and the idea of electromagnetic disturbances propagating at most at the speed of light. The resolution of this conflict lies in the fact that electromagnetic disturbances cannot be specified freely but are subject to

[2]Supported by Fonds zur Förderung der wissenschaftlichen Forschung in Österreich, Project No. P9376–PHY.

constraints. More precisely, the Maxwell equations consist of two sets: the hyperbolic system of evolution equations, and an elliptic system – the constraint equations. (In elementary treatises of electromagnetic theory these two aspects of the theory are hidden due to the almost ubiquitous use of retarded fields.) To appreciate this point consider first a theory with no constraints, namely that of a massless scalar field Φ on Minkowski spacetime, i.e.

$$\Box\Phi = h^{\mu\nu}\nabla_\mu\nabla_\nu\Phi = \rho \qquad (1.1)$$

with $h_{\mu\nu}$ the Minkowski metric on $\mathbf{R}^4$, ρ a source of compact support in space together with, perhaps, some equation governing the evolution of ρ in time. Let $x^\mu = (t, x^a)$ be standard Minkowskian coordinates. Then we are free to choose

$$f(x^a) = \Phi(t, x^a)|_{t=0} \qquad \text{and} \qquad g(x^a) = \frac{\partial}{\partial t}\Phi(t, x^a)|_{t=0},$$

e.g. we could take $f \equiv g \equiv 0$. If, for example, the support of ρ consists of two disjoint sets in x^a–space diffeomorphic to solid balls in $\mathbf{R}^3$, i.e. two spherical "bodies", these bodies would, for $t = 0$, have no interaction at all and only feel each others influence after the time it takes light to cross the empty space which separates them.

Take, next, electromagnetism. The Maxwell equations can be written as a hyperbolic system for the electric field $\vec{E}$ and the magnetic field $\vec{B}$. Let us concentrate on $\vec{E}(t, \vec{x})$, $\vec{x} = (x^a)$. $\vec{E}\big|_{t=0}$ is not arbitrary, but satifies

$$\vec{\nabla}\vec{E} = 4\pi\rho \qquad \text{at } t = 0, \qquad (1.2)$$

where ρ is the charge density. When ρ is not identically zero, clearly $\vec{E}$ cannot be identically zero either. But suppose that ρ, as before, is supported in two compact, disjoint "islands" $\mathcal{B}_i$ in space. Can we at least have $\vec{E} = 0$ (i.e. no interaction between the connected components of supp ρ) outside of supp ρ? The answer is "yes", provided that the charges Q_i $(i = 1, 2)$, given by

$$Q_i = \int_{B_i \subset \mathbf{R}^3} \rho d^3x = \frac{1}{4\pi} \int_{\partial B_i} \vec{E} d\vec{S} \tag{1.3}$$

are zero.

To prove the first statement, apply the Helmholtz–Hodge decomposition on $\vec{E}$, giving

$$\vec{E} = \vec{\nabla}\varphi + \vec{\nabla} \times \vec{\psi}. \tag{1.4}$$

When $(r := |\vec{x}|)$ $\vec{E} = O(1/r^2)$, $\partial \vec{E} = O(1/r^3)$, a.s.o., in short: $\vec{E} = O^\infty(1/r^2)$, there is always a scalar field φ and a vector field $\vec{\psi}$, both $O^\infty(1/r)$, such that (1.4) is valid. Inserting (1.4) into (1.2), we obtain

$$\Delta\varphi = 4\pi\rho. \tag{1.5}$$

Equ. (1.5) has a unique solution $\varphi = O^\infty(1/r)$. We now choose $\vec{\psi}$, such that

$$\vec{\nabla} \times \vec{\psi} = -\vec{\nabla}\varphi \text{ outside supp } \rho. \tag{1.6}$$

This, by the de Rham theory, is possible iff Q_1 and Q_2 both vanish. This argument also shows that φ describes the "Coulombic" aspect of $\vec{E}$ whereas $\vec{\psi}$ modulo closed 1–forms plays the role of "free data" or "radiative excitations".

The above considerations merely serve the purpose of providing a simple model for what goes on in General Relativity. Here, again the field equations can be turned into a set of evolution equations, supplemented by the underdetermined elliptic system of constraint equations any initial data has to satisfy. Again one can identify a set of free data which, due to the nonlinear nature of the theory, is much more complicated and a long–range aspect. The role of the charge Q in electromagnetism is now played by the ADM energy m. But there is a crucial difference. In the absence of sources, the total charge Q of a Maxwell field is zero. In contrast for reasonable matter and in particular in vacuo, the quantity m is always ≥ 0 and m is zero iff the data are ones for Minkowski spacetime [14]. Thus the gravitational field has always a long–range component except in the physically trivial situation.

The present lecture, based on joint work with Niall Ó Murchadha, is devoted to a study of the set of solutions of the Einstein vacuum constraints in the asymptotically flat case. The most interesting properties of this set are the ones relevant for the global Cauchy evolution. When the initial data are uniformly close to ones for Minkowski spacetime we know, thanks to the work of Christodoulou and Klainerman [8], that the time–evolved spacetime remains close to Minkowski space. In particular the maximal Cauchy development is geodesically complete. On the other hand, it follows from a by–now classic result of Penrose [12], that an asymptotically flat spacetime which is globally hyperbolic (i.e. results from a Cauchy problem) and has a trapped surface, is null geodesically incomplete. Thus our interest will be in initial data for the Einstein vacuum equations containing trapped surfaces.

2 The Hamiltonian Constraint of G.R. and Outer Trapped Surfaces

The Einstein vacuum equations are

$$R_{\mu\nu}[h] = 0, \tag{2.1}$$

where $h_{\mu\nu}$ is a Lorentz metric on a 4–manifold N and $R_{\mu\nu}$ its Ricci tensor. Let $M \subset N$ be a submanifold which is spacelike with respect to $h_{\mu\nu}$. Let g_{ab} be the Riemannian metric on M induced by h and $p_{ab} = p_{(ab)}$ be the extrinsic curvature. The object g plays the role of the configureation space variable (analogous to $\Phi(0, x)$ in the scalar field example of § 1) and p_{ab} the momentum space variable (analogous to $\frac{\partial}{\partial t}\Phi(0, x)$). By virtue of Equ. (2.1), (g, p) have to satisfy the following system of PDE's:

$$\mathcal{R}[g] + p^2 - p_{ab}p^{ab} = 0 \tag{2.2}$$

$$D^a(p_{ab} - pg_{ab}) = 0, \tag{2.3}$$

i.e. the vacuum constraints on M. Here $\mathcal{R}$ is the scalar curvature, $p := g^{ab}p_{ab}$ and all indices are raised by means of g^{ab}. There are two classes of situations for which (2.2,3) have been studied extensively: a) the case where M is compact–without–boundary, sometimes simply called the cosmological case, and b) the case where (M, g_{ab}, p_{ab}) are asymptotically flat. We focus on b). For ease of presentation we shall take M to be diffeomorphic to $\mathbf{R}^3$. Our preliminary definition of asymptotic flatness is that there exist coordinates x^a, such that

$$g_{ab} - \delta_{ab} = O\left(\frac{1}{r}\right), \qquad p_{ab} = O^{\infty}\left(\frac{1}{r^2}\right). \tag{2.4}$$

We first consider the special case, in which $p_{ab} = 0$ ("momentarily staric" initial data) where (2.3) is trivially satisfied, and start by malting a small digression on conformal rescalings. Let $\widetilde{M}$ be diffeomorphic to S^3 with $\widetilde{g}_{ab}$ the standard metric theorem and Λ be a point $\in \widetilde{M}$. There exists a positive function G, smooth on $\widetilde{M} \setminus \Lambda$, such that $G^4\widetilde{g}_{ab}$ is the flat metric on $\widetilde{M} \setminus \Lambda \cong \mathbf{R}^3$. Of course, G is nothing but the conformal factor arising from stereographic projection. G blows up at Λ like

$$G = \frac{1}{|\widetilde{x}|} + O(|\widetilde{x}|), \tag{2.5}$$

where $|\widetilde{x}|$ is the geodesic distance from Λ w.r. to $\widetilde{g}_{ab}$. We will find it convenient to define asymptotic flatness of a metric g_{ab} on M by the possibility of obtaining g_{ab} from a conformal decompactification siinilar to ste·eographic projection which sends Λ to infinity. There is yet another role of conformal rescalings in the present context, which arises as follows: To solve

$$\mathcal{R}[g] = 0, \tag{2.6}$$

i.e. Equ. (2.2) when $p_{ab} = 0$, we choose a "background" metric $\widetilde{g}_{ab}$ and try to find $G > 0$, such that $g_{ab} = G^4 \widetilde{g}_{ab}$ satisfies (2.6). Let us recall a basic formula. Define the conformal Laplacian L_g by

$$L_g := -\Delta_g + \frac{1}{8}\mathcal{R}[g], \tag{2.7}$$

where $\Delta_g = g^{ab} D_a D_b$. Then

$$L_g \Phi = G^{-5} L_{\widetilde{g}} \widetilde{\Phi}, \qquad G > 0, \tag{2.8}$$

where $\widetilde{\Phi}$ is a scalar function and

$$\Phi = G^{-1} \widetilde{\Phi}. \tag{2.9}$$

Putting $\widetilde{\Phi} = G$, so that $\Phi \equiv 1$, whence $\Delta_g \Phi = 0$, we find that

$$L_g \Phi = \frac{1}{8}\mathcal{R}[g] = G^{-5} L_{\widetilde{g}} G. \tag{2.10}$$

Thus g solves (2.6), iff

$$L_{\widetilde{g}} G = 0. \tag{2.11}$$

We now combine the above–mentioned roles of conformal rescalings. We pick a smooth metric $\widetilde{g}_{ab}$ on $\widetilde{M} \cong S^3$ and try to solve

$$L_{\widetilde{g}} G = 4\pi \delta|_\Lambda , \tag{2.12}$$

for $G > 0$ on $\widetilde{M} \setminus \Lambda$, where $\delta|_\Lambda$ is the delta distribution concentrated at Λ. Equ. (2.12) can always be solved locally, i.e. in a neighbourhood of Λ (see e.g. [9]). Any solution obeys the estimate

$$G = \frac{1}{|\widetilde{x}|} + m + O^\infty(|\widetilde{x}|), \tag{2.13}$$

where m is a constant. It is now easy to see that the "physical" metric $g = G^4 \tilde{g}$ satisfies

$$g_{ab} = \delta_{ab} \left(1 + \frac{m}{2r'}\right) + O^\infty \left(\frac{1}{r'^2}\right) \tag{2.14}$$

in "Kelvin–transformed" coordinates $x'^a = \tilde{x}^a/|\tilde{x}|^2$, $\tilde{x}^a$ being Riemannian normal coordinates of $\tilde{g}_{ab}$ centered at Λ and $r = (x^a x^b \delta_{ab})^{1/2}$. Thus the constant m is the ADM mass of the initial data. Suppose a global solution $G > 0$ exists. Then this is unique. Since, suppose, there were two solutions, G and G': then the function Φ on M, defined by $\Phi = G'G^{-1}$ satisfies

$$\Delta_{\tilde{g}}\Phi = 0, \qquad \Phi \to 1 \text{ at } \Lambda \tag{2.15}$$

and is thus, by the maximum principle, identically equal to one. Furthermore, let $\tilde{g}'_{ab}$ be another metric on $\widetilde{M}$, conformal to $\tilde{g}_{ab}$, i.e. $\tilde{g}'_{ab} = \omega^2 \tilde{g}_{ab}$, $\omega > 0$. Then

$$G' = C^{-1/2}\omega^{-1/2}G, \tag{2.16}$$

with $C := \omega|_\Lambda$ satisfies (2.12) with $\tilde{g}$ replaced by $\tilde{g}'$ and

$$g'_{ab} = C^{-2}g_{ab} \tag{2.17}$$

and the physical spacetime metric evolving from these data satisfies

$$h'_{\mu\nu} = C^{-2}h_{\mu\nu}. \tag{2.18}$$

Thus a conformal rescaling of $\tilde{g}_{ab}$ merely results in a constant rescaling of $h_{\mu\nu}$, which is in turn a rescaling of the mass, i.e.

$$m' = C^{-1}m. \tag{2.19}$$

The conformal geometry associated with $\tilde{g}_{ab}$ is the analogue of the vector field $\vec{\psi}$ in the Maxwell example of the previous section, G is the analogue of the scalar field φ, and, as pointed out, m plays a role similar to the electric charge Q. Suppose, now, that the "radiative excitations have compact support", i.e. the metric $\widetilde{g}_{ab}$ is (conformally) flat mear Λ. Does it follow that G near Λ is close to the trivial one discussed previously? The answer is "no": Except in the case where $\widetilde{g}_{ab}$ is everywhere conformal to the standard metric on S^3, G will, by virtue of the positive–mass theorem [14], not satisfy the estimate (2.6), since $m > 0$.

We now come to the question as to when Equ. (2.12) has a solution $G > 0$. The answer is as follows. The differential operator $L_{\widetilde{g}}$ can be viewed as an essentially self–adjoint operator on the Hilbert space of square–integrable functions on $\widetilde{M}$ with volume element $dV_{\widetilde{g}}$. This operator known to have a pure point spectrum, which is bounded from below. The lowest eigenvalue $\lambda_1(\widetilde{g})$ is non–degenerate, the lowest eigenfunction $f_1(\widetilde{g})$ nowhere zero on $\widetilde{M}$. Suppose $\lambda_1(\widetilde{g})$ was zero. Then, clearly, (2.12) has no solution G at all, positive or otherwise. But one could use f_1 as conformal factor. The metric $f_1^4 \widetilde{g}_{ab}$ will now furnish a smooth metric on S^3 satisfying (2.6). We would thus have a cosmological solution of the time–symmetric constraints. The basic existence result, essentially found in [5], is this

Theorem: $G > 0$ solving (2.12) exists, if and only if $\lambda_1(\widetilde{g}) > 0$.

We now introduce the notion of critical sequences (CS's). These are sequences $\widetilde{g}_n$ of metrics ($n \in \mathbf{N}$) on $\widetilde{M}$, all of which have $\lambda_1(\widetilde{g}_n) > 0$ and which tend, say, uniformly with all derivatives, to a metric $\widetilde{g}_\infty$ having $\lambda_1(\widetilde{g}_\infty) = 0$. To see that such sequences exist, we use the following facts. Firstly, there exist plenty of metrics $\widetilde{g}$ on $\widetilde{M}$ having $\lambda_1(\widetilde{g}) > 0$. E.g., the standard metric on S^3 has this property and, since $\lambda_1(\widetilde{g})$ is continuous in the C^2–topology [10], any other metric sufficiently :lose in the latter topology will also have $\lambda_1(\widetilde{g}) > 0$. Furthermore it is well known [4] that any compact manifold has

Riemannian metrics $\widetilde{h}$ having $\lambda_1(\widetilde{h}) < 0$. Now consider the metrics $\widetilde{g}_t = t\widetilde{g} + (1-t)\widetilde{h}$ for $t \in [0,1]$. Since $\lambda_1(\widetilde{g}_0)$ is positive and $\lambda_1(\widetilde{g}_1)$ is negative, there exists, again by continuity of λ_1, a number $s \in (0,1)$, such that $\lambda_1(\widetilde{g}_t) > 0$ for $t \in [0,s)$ and $\lambda_1(\widetilde{g}_s) = 0$. Thus $\widetilde{g}_n = \widetilde{g}_{s(1-1/n)}$ is a CS. For explicit (numerical) examples, see [2].

There is the following

Theorem [1]:

(i) The ADM masses m_n along a CS go to infinity.

(ii)

$$G_n = \frac{1}{|\widetilde{x}|} + \frac{m_n}{2} + m_n O^\infty(|x|) \tag{2.20}$$

$$\partial\left(G_n - \frac{1}{|\widetilde{x}|}\right) = m_n O^\infty(1). \tag{2.21}$$

In (2.20,21) $|\widetilde{x}|$ is taken to be a suitable approximation to the geodesic distance from Λ. It is understood that the constants involved in O^∞ in (2.20,21) can be taken to be independent of n. If $\widetilde{g}_n$ and $\partial\widetilde{g}_n$ all coincide at Λ, $|\widetilde{x}|$ can be taken to be independent of n. Note that (i) entails an independent proof of positivity of mass for large n. The estimates (2.20,21) are the expression of the fact that "space closes up" as n tends to infinity. We now compute H_n, the mean curvature with respect to the physical metric $g_n = G^4 \widetilde{g}_n$ of the spheres $|\widetilde{x}| = R$ for small R. We conclude from the previous result that

$$H_n = \frac{4 G_n^{-3}}{|\widetilde{x}|(\widetilde{D}^a |\widetilde{x}|^2 \widetilde{D}_a |\widetilde{x}|^2)^{1/2}} \left[1 - \frac{m_n |\widetilde{x}|}{2} + m_n O(|\widetilde{x}|^2)\right] \tag{2.22}$$

from which we infer the

Theorem: There exist constants ε and δ independent of n, such that the surfaces $|\widetilde{x}| = R$ with

$$\varepsilon \leq \frac{1}{R} \leq \frac{m_n}{2} - \delta \tag{2.23}$$

have $H_n < 0$.

The spacetime meaning of $H_n < 0$ is that null geodesics normal to $|\widetilde{x}| = R$ which are outgoing, i.e. pointing towards Λ, initially converge. Thus $|\widetilde{x}| = R$ are outer–trapped surfaces, provided R satisfies (2.23) and n is large. Thus, by (a slight generalization of [15]) the Penrose singularity theorem, any Cauchy evolution (N, h) of (M, g) is null geodesically incomplete. In fact, by time–symmetry this conclusion holds both for the future and the past. Moreover one would expect that trapped surfaces will exist on all Cauchy surface . So this is not the sort of situation one has in mind when thinking of a gravitational collapse. Rather one would envisage a time–asymmetric situation, where an apparently innocuous space-time develops future trapped surfaces in the course of time. Future trapped surfaces are compact spacelike surfaces for which both the outgoing and the ingoing null normals initially converge. The surfaces constructed above are not at all like this: they are outer–trapped and inner anti–trapped (as opposed to compact spacelike 2–surfaces in Minkowski space, which are inner trapped and outer anti–trapped). To produce future trapped surfaces, we have to put extrinsic curvature into our initial data sets, which we do in the next section.

3 Momentum Constraints with $p = 0$

Most work done so far on (2.2,3) deals with the maximal case where $p = 0$, so that

$$\mathcal{R}[g] = p_{ab}p^{ab} \tag{3.1}$$

$$D^a p_{ab} = 0, \qquad p = 0. \tag{3.2}$$

Equ.'s (3.2) are conformally invariant in the following sense: When g_{ab} solves (3.2) with respect to the metric g_{ab}, $\bar{p}_{ab} = \omega^{-1} p_{ab}$ ($\omega > 0$) does so relative to $\bar{g}_{ab} = \omega^2 g_{ab}$. Thus one could solve $\tilde{D}^a \tilde{p}_{ab} = 0$, $\tilde{p} = 0$ on the unphysical manifold $(\widetilde{M}, \tilde{g})$ and obtain the physical p_{ab} as $p_{ab} = G^{-2} \tilde{p}_{ab}$ where, now, G has to satisfy

$$L_{\tilde{g}} G = 4\pi \delta|_{\Lambda} + \frac{1}{8} G^{-7} \tilde{p}_{ab} \tilde{p}^{ab}, \tag{3.3}$$

so that

$$\frac{1}{8} \mathcal{R}[g] = G^{-5} L_{\tilde{g}} G = \frac{1}{8} G^{-12} G^4 G^8 p_{ab} p^{ab}, \tag{3.4}$$

i.e. $\mathcal{R}[g] = p_{ab} p^{ab}$ on M. However we prefer a somewhat roundabout procedure where a certain "background" TT–tensor p'_{ab} is first found on M with a background metric g'_{ab}, and only then Equ. (3.3) solved. Let Ω be a smooth, positive function on $\widetilde{M}$, satisfying

$$\Omega|_{\Lambda} = 0, \qquad \tilde{D}_a \Omega\Big|_{\Lambda} = 0, \qquad \left(\tilde{D}_a \tilde{D}_b \Omega - 2\tilde{g}_{ab}\right)\Big|_{\Lambda} = 0, \tag{3.5}$$

i.e. Ω approximates $|\tilde{x}|^2$ of the previous section. Then

$$g'_{ab} = \Omega^{-2} g_{ab} \tag{3.6}$$

is an asymptotically flat metric on M with vanishing mass, i.e.

$$g'_{ab} - \delta_{ab} = O^{\infty}\left(\frac{1}{r^2}\right). \tag{3.7}$$

We solve

$$D'^a p'_{ab} = 0, \qquad p' = 0, \tag{3.8}$$

using the standard York–method [7]. We shall for the remainder of this section for typographical reasons omit the prime from differential operators with the understanding that these operations are meant to be with respect to the metric g'_{ab} rather than g_{ab}. Pick a tracefree tensor Q_{ab} and try to find a covector field W_a so that p'_{ab} given by

$$Q_{ab} = (LW)_{ab} + p'_{ab} \tag{3.9}$$

is TT. Here L is defined by

$$(LW)_{ab} = D_a W_b + D_b W_a - \frac{2}{3} g'_{ab} D^c W_c. \tag{3.10}$$

Thus we have to solve

$$\Delta W_a + \frac{1}{3} D_a(D^b W_b) + R'^{\ b}_a W_b = D^a Q_{ab}. \tag{3.11}$$

When Q_{ab} is $O^\infty(1/r^{2+\epsilon})$, $\epsilon > 0$, a unique W_a exists which is $O^\infty(1/r^2)$ [6]. We shall require the faster fall–off

$$Q_{ab} = O^\infty\left(\frac{1}{r^{3+\epsilon}}\right), \qquad \epsilon > 0. \tag{3.12}$$

Then the results in [6] can be improved to give

$$W_a = \overset{1}{W_a} + \overset{2}{W_a} + \overset{3}{W_a} + \overset{4}{W_a} + O^\infty\left(\frac{1}{r^{2+\epsilon}}\right) \tag{3.13}$$

with

$$
\begin{aligned}
\overset{1}{W_a} &= \frac{7}{8}\frac{P_a}{r} + \frac{1}{8}\frac{x_a P_b x^b}{r^3} \\
\overset{2}{W_a} &= \frac{3}{4}\frac{L_{ab} x^b}{r^3} \\
\overset{3}{W_a} &= \frac{1}{4} C \partial_a \frac{1}{r} \\
\overset{4}{W_a} &= \frac{3}{4}\frac{M_{ab} x^b}{r^3} + \frac{3}{8}\frac{x_a M_{bc} x^b x^c}{r^5}.
\end{aligned}
\tag{3.14}
$$

All indices are lowered and raised with δ_{ab}. Furthermore P_a, $L_{ab} = L_{[ab]}$, C, $M_{ab} = M_{(ab)}$ with $M_{ab}\delta^{ab} = 0$ are all constants. The term $\overset{1}{W_a}$ is of order $1/r$, the other ones are of order $1/r^2$. P_a and L_{ab} are respectively proportional to the ADM 3–momentum and angular momentum both of which are conserved under time evolution. Neither the scalar constant C, nor the quadrupole–type quantity M_{ab} are conserved. Of particular interest for us is $\overset{3}{W}_a$. It's contribution to p_{ab} is given by

$$(L \overset{3}{W})_{ab} = -\frac{3C}{4} \frac{x_a x_b - \frac{1}{3}\delta_{ab}r^2}{r^5} + O^\infty\left(\frac{1}{r^5}\right) \qquad (3.15)$$

which is asymptotically spherically symmetric. In particular $(L \overset{3}{W})_{ab}x^a x^b$ has, asymptotically, a sign, namely $-C$, whereas none of the other terms in (3.14) have this property. Since this will be crucial in our trapped–surface proof we will need the existence of "source tensors" Q_{ab}, such that P_a, L_{ab} and M_{ab} are all zero and C is non–zero. That this is possible, is a special case of a result proved in [3]. Namely, we have the

Theorem: When g'_{ab} is not conformally flat, there exists Q_{ab} such that W_a satisfying Equ. (3.11) agrees with $\overset{3}{W}_a$ to leading order.

We sketch the proof. Consider covector fields λ_a satisfying

$$\Delta\lambda_a + \frac{1}{3}D_a(D^b\lambda_b) + \mathcal{R}'^{\,b}_a W_b = 0, \qquad (3.16)$$

i.e. the homogeneous form of (3.11), but blowing up at infinity as follows

$$
\begin{aligned}
\overset{1}{\lambda}_a &= \beta_a + O\left(\frac{1}{r^\varepsilon}\right) \\[4pt]
\overset{2}{\lambda}_a &= \beta_{ab}x^b + O\left(\frac{1}{r^\varepsilon}\right), \qquad \beta_{ab} = \beta_{[ab]} \\[4pt]
\overset{3}{\lambda}_a &= \frac{\mu}{3}x_a + O\left(\frac{1}{r^\varepsilon}\right) \\[4pt]
\overset{4}{\lambda}_a &= \mu_{ab}x^b + O\left(\frac{1}{r^\varepsilon}\right), \qquad \mu_{ab} = \mu_{(ab)}, \qquad \mu_{ab}\delta^{ab} = 0,
\end{aligned}
\qquad (3.17)
$$

where β_a, β_{ab}, μ and μ_{ab} are all constants. These exist, as can be inferred from [6]. Next define

$$\langle \lambda | W \rangle = \frac{1}{4\pi} \oint_{r'=\infty} [(LW)_{ab}\lambda^b - (L'\lambda)_{ab}W^b]d\Sigma^a. \qquad (3.18)$$

A computation yields

$$
\begin{aligned}
\langle \overset{1}{\lambda} | W \rangle &= P_a\beta^a \\
\langle \overset{2}{\lambda} | W \rangle &= \frac{3}{4}L_{ab}\lambda^{ab} \\
\langle \overset{3}{\lambda} | W \rangle &= \frac{1}{2}C\mu \\
\langle \overset{4}{\lambda} | W \rangle &= \frac{11}{10}M_{ab}\mu^{ab}.
\end{aligned}
\qquad (3.19)
$$

On the other hand we have that

$$
\begin{aligned}
\langle \lambda | W \rangle &= -\int_M D^a(LW)_{ab}\lambda^b dV_{g'} = \\
&= -\int_M (D^a Q_{ab})\lambda^b dV_{g'} = \\
&= \int_M Q^{ab}(L\lambda)_{ab}dV_{g'}.
\end{aligned}
\qquad (3.20)
$$

Let A run through the 12–parameter set of λ's according to (3.17). Set

$$\overset{A}{Q}_{ab} = \sigma(L \overset{A}{\lambda})_{ab} \qquad (3.21)$$

where σ is a smooth, positive function of $O(1/r^{4+\varepsilon})$. We try to find $\overset{A}{C}$, such that $\sum_A \overset{A}{C}\overset{A}{Q}_{ab}$ gives rise to prescribed moments $\overset{A}{M}$ in W_a. Using the pairing between the λ's and the W's given by (3.19) and using (3.21), this requires solving the equation

$$\overset{A}{M} = \sum_B \overset{B}{C} f^{AB}, \qquad (3.22)$$

where

$$f^{AB} = f^{(AB)} = \int_M \sigma(L \overset{A}{\lambda})_{ab}(L \overset{B}{\lambda})^{ab}dV_{g'}. \qquad (3.23)$$

Equ. (3.22) is a linear equation in 12 unknowns. By symmetry of f^{AB} it can be solved iff $\overset{A}{M}$ is orthogonal to the null space of f^{AB}. The null space of f^{AB}, by (3.23), consists precisely of those vectors d^A, such that $\lambda = \sum_A d^A \overset{A}{\lambda}$ has $(L\lambda)_{ab} = 0$, i.e. is a conformal Killing vector (CKV). This immediately requires that $\mu_{ab} = 0$. Furthermore, since g'_{ab} is not conformally flat, we also have to have $\mu = 0$. Because suppose μ was $\neq 0$. The vector field λ^a extends to a CKV on the conformally compactified manifold $(\widetilde{M}, \widetilde{g})$. But it is easy to see that this CKV would have to be "essential", i.e. cannot be made a Killing field for a metric in the conformal class of $\widetilde{g}_{ab}$. But, by a theorem due to Obata [11], this would imply that $\widetilde{g}_{ab}$ is conformal to the standard metric on S^3. It follows that the vector $\overset{A}{M}$, given by $P_a = 0$, $L_{ab} = 0$, $C \neq 0$, $M_{ab} = 0$ is orthogonal to the null space of f^{AB}, which ends the proof.

To solve the full set of maximal constraints we do the following. First pick a metric $\widetilde{g}_{ab}$ on $\widetilde{M}$ satisfying $\lambda_1(\widetilde{g}) > 0$. Then decompactify using the conformal factor Ω as in (3.5), pick a trace–free tensor Q_{ab} and solve (3.11). From this obtain $\widetilde{p}_{ab} = \Omega^{-1} p'_{ab}$ where p'_{ab} is given by (3.9). It remains to solve (3.3) for $G > 0$. This can be done [1], when $\lambda_1(\widetilde{g}) > 0$. Then the physical initial data set (M, g_{ab}, p_{ab}) is given by

$$g_{ab} = G^4 \widetilde{g}_{ab}, \qquad p_{ab} = G^{-2} \widetilde{p}_{ab}. \qquad (3.24)$$

4 Trapped Surfaces for Maximal Data

For a 2–surface on a maximal initial data set (M, g_{ab}, p_{ab}) the divergence of the outgoing (resp. ingoing) null normal is given by

$$\Theta_{\text{out,in}} = \pm H + p_{ab} n^a n^b \qquad (4.1)$$

where H is the mean curvature of the surface in M and n^a it's normal, pointing into the "out"–direction. For a first orientation it is useful to look at spherically symmetric, maximal slices of the Kruskal spacetime. There are three types of such slices (see e.g. [13]). Firstly there are the time–symmetric Cauchy surfaces $t = t_0$. These have a minimal surface at $r = 2m$ ((t, r) are standard Kruskal coordinates), where $\Theta_{\text{out,in}} = \pm H$ changes sign. In fact, in spacetime terms these minimal surfaces are all the same, namely the bifurcation surface of the event horizon. The trapped surfaces constructed in Section 2 are in a sense modelled on what happens in Kruskal on such a time–symmetric slice, although this model should not be taken too literally for two reasons. Firstly, in the Kruskal spacetime $r = r_0$ separates two asymptotically flat regions in $t = t_0$ whereas the surfaces we studied have an asymptotically flat exterior and a compact interior (in particular, the Penrose singularity theorem strictly speaking does not apply to $t = t_0$–slices of Kruskal). Secondly, the vacuum data we have constructed are never spherically symmetric.

The remaining classes of spherically symmetric maximal slices have $p_{ab} \neq 0$ and, in particular, a non–vanishing term near infinity involving C as in (3.14). There are two possibilities. Either $C^2 < 27m^4$: These enter, say, the future horizon at $r = 2m$, reach a minimum value of r (at a minimal surface) for some r_0 with $2m > r_0 > 3m/2$ and leave the black hole region through the other branch of the future horizon. All spheres for $r_0 < 2m$ are future trapped. The third class of slices is given by $C^2 > \frac{27}{16}m^4$: These have no minimal surfaces, but rather they hit the singularity $r = 0$. (In the limiting case $C^2 = \frac{27}{16}m^4$ the maximal slice approaches $r = 3m/2$ near the other branch of the future horizon.)

Going back to the general case, $\Theta_{\text{out,in}}$ for the 2–spheres $\Omega = \Omega_0 = $

constant is given by

$$\Theta_{\text{out,in}} = \frac{4G^{-3}\Omega^{-1/2}}{(\Omega_c\Omega^c)^{1/2}}[\pm A - B] \tag{4.2}$$

where

$$A = \frac{1}{4}\Omega^{1/2}G\left(\widetilde{g}^{ab} - \frac{\Omega^a\Omega^b}{\Omega_c\Omega^c}\right)\Omega_{ab} + \Omega^{1/2}\Omega^a G_a \tag{4.3}$$

$$B = \frac{1}{4}\frac{G^{-3}\Omega^{1/2}}{(\Omega_c\Omega^c)^{1/2}}\widetilde{p}_{ab}\Omega^a\Omega^b. \tag{4.4}$$

$\widetilde{p}_{ab}$ enters the expression for Θ in 2 ways: explicitly in (4.4) and implicitly in (4.3) and (4.4) through the dependence of G on $\widetilde{p}_{ab}$ by virtue of (3.3). Evaluating B at the point Λ, when p'_{ab} has just the C–term to leading order, we find that

$$B = \frac{C}{4}\Omega + O^\infty(\Omega^{1+\epsilon/2}). \tag{4.5}$$

Furthermore $A = 1 + O^\infty(\Omega^{1/2})$. Trapped surfaces are ones on which Θ_{out} and Θ_{in} are both everywhere negative. Thus one is led to look for sequences $(\widetilde{g}_n, \widetilde{p}_n)$ for which B dominates A as $n \to \infty$ for some range of Ω's. For example one could try to fix $\widetilde{g}_n$ and let C become large. However heuristic scaling arguments[3] indicate that, in this case A and B are of the same order of magnitude for small Ω. The other possibility is to take again a critical sequence $\widetilde{g}_n$. By the definition of criticality $\widetilde{g}_n$ can not be conformally trivial for large n. We can thus take, for each g'_n, a source tensor Q_{ab} giving us a p'_{ab} with just the C–term to leading order. It is easy to see that this can be done in such a manner that the error–terms in p'_{ab} are independent of n. We now choose $\Omega^{1/2} = \Omega_0^{1/2} = 2/m_n +$ correction–terms in such a manner that $(m_n \to \infty$ as $n \to \infty)$

$$A = 0 + O(m_n^{-3}) \tag{4.6}$$

---

[3]I thank Robert Bartnik for advice on how to do this.

and

$$B = C + O(m_n^{-3}). \tag{4.7}$$

Thus $\Theta_{\text{out,in}} < 0$ as $n \to \infty$ for $\Omega = \Omega_0$. A more precise statement and detailed proof of this result will be given in forthcoming work by the author and N. Ó Murchadha. It should be clear that the trapped surfaces we construct are modelled on the ones occurring in the second class of maximal slices in the Kruskal spacetime discussed previously.

As opposed to our result on outer trapped surfaces, the (physical) volume of the region in which we are able to prove existence of trapped surfaces does not go to infinity as $n \to \infty$, but goes to zero. To see whether this can be improved would require a more detailed control of the nonlinearities in (3.3).

Let us end with some remarks. The critical sequences of initial data we have considered in this work can for large n not result from Cauchy evolution of one of them since $m_n \to \infty$ whereas the ADM mass would have to be conserved. Thus our results do not prove that trapped surfaces can emerge by time evolving innocuous, i.e. trapped–surface–free data. Still one expects this to be true. It would be interesting to find a proof. Much more ambitiously, one would like to understand the nature of the final singularities of the maximal Cauchy evolution of the data we have found. This issue, which touches on the question of cosmic censorship, is of course wide open.

Acknowledgments

I thank Robert Bartnik for helpful discussions.

References

[1] R. Beig and N. Ó Murchadha, "Trapped Surfaces Due to Concentration of Gravitational Radiation", *Phys. Rev. Lett.* **66** (1991) 2421–2424. For an extension of these results to outer trapped surfaces to maximal data see R. Beig and N. Ó Murchadha, "Trapped surfaces in vacuum spacetimes", *Class. Quantum Grav.* **11** (1994) 419–430.

[2] R. Beig and S. Husa, "Initial Data for General Relativity with Toroidal Conformal Symmetry", *Phys. Rev. D* **50** (1994) R7116–R7118 and S. Husa, in preparation.

[3] R. Beig and N. Ó Murchadha, "The Momentum Constraints of General Relativity and Spatial Conformal Isometries", (1995) gr–qc 9412029.

[4] L. Bérard Bergery, "Scalar Curvature and Isometry Groups" in: *Spectra of Riemannian Manifolds*, ed. by M. Berger et al., Kagai, Tokyo (1993).

[5] M. Cantor and D. Brill, "The Laplacian on Asymptotically Flat Manifolds and the Specification of Scalar Curvature", *Compositio Math.* **43** (1981) 317–330.

[6] A. Chaljub–Simon, "Decomposition of Covariant Two–Tensors on $\mathbf{R}^3$", *General Rel. Grav.* **14** (1982) 743–749.

[7] Y. Choquet–Bruhat and J. W. York, "The Cauchy Problem," in *General Relativity and Gravitation*, Vol. 1, ed. by A. Held, Plenum, New York (1980)

[8] D. Christodoulou and S Klainerman, *The Global Nonlinear Stability of Minkowski Space*, Princeton University Press, Princeton (1993).

[9] P. R. Garabedian, *Partial Differential Equations*, Wiley, New York (1964).

[10] T. Kato, *Perturbation Theory for Linear Operators*, Springer, New York (1966).

[11] M. Obata, "The Conjectures on Conformal Transformations of Riemannian Manifolds", *J. Diff. Geom.* **6** 1971 237–258.

[12] R. Penrose, "Gravitational Collapse and Space–Time Singularities", *Phys. Rev. Lett.* **14** (1965) 57–59.

[13] B. L. Reinhart, "Maximal Foliations of Extended Schwarzschild Space", *J. Math. Phys.* **14** (1973) 719.

[14] R. Schoen and S-T. Yau, "On the Proof of the Positive Mass Conjecture", *Commun. Math. Phys.* **65** (1979) 45–76 and E. Witten, "A New Proof of the Positive Energy Theorem", *Commun. Math. Phys.* **80** (1981) 381–402.

[15] G. Totschnig, *Ein Singularitätentheorem für Outer Trapped Surfaces*, diploma thesis, University of Vienna (1994).

Part II:
Global Structures

Black Hole Collisions, Analytic Continuation and Cosmic Censorship

Dieter R. Brill[1]

[1]Department of Physics, University of Maryland, College Park, MD 20742, USA

Abstract.   Exact solutions of the Einstein-Maxwell equations that describe moving black holes in a cosmological setting are discussed with the aim of discovering the global structure and testing cosmic censorship. Continuation beyond the horizons present in these solutions is necessary in order to identify the global structure. Therefore the possibilities and methods of analytic extension of geometries are briefly reviewed. The global structure of the Reissner-Nordström-de Sitter geometry is found by these methods. When several black holes are present, the exact solution is no longer everywhere analytic, but less smooth extensions satisfying the Einstein equations everywhere are possible. Some of these provide counterexamples to cosmic censorship.

1   Introduction

The recently discovered [1] exact *dynamical* solution of Einstein's equations provides a common thread that ties together three items of the title. This solution describes a cosmology with several charged black holes in motion and capable of collision. To discover the details of the collision one needs to continue the solution beyond the region in which it was originally defined. The spacetime so continued can then contain a naked singularity and provide a counterexample to some versions of the cosmic censorship hypothesis.

Because interesting applications of Einstein's equations frequently occur in manifolds of complicated topology, which cannot be covered by a single coordinate patch, the region in which a typical exact solution is first known is often incomplete. The simplest way to complete it, if possible, is by analytic extension. It is remarkable how little is known in a systematic way about this frequently encountered problem. Here we will not materially improve on this situation, but merely recall some of what is known about analytic continuation, and discuss the most common class of geometries for which a method exists.

Although it is tempting to expect that such a method can also analytically continue the colliding black holes of interest here, we will find that, surprisingly, these holes are in general not everywhere analytic. There is of course nothing unphysical about such behavior; it is what one expects in the presence of a gravitational waves pulse. We exhibit continuations that are as smooth as possible across the associated Cauchy horizon. The naked singularities of interest for cosmic censorship are then found on the other side of such horizons. (Other horizons are cosmological and do not hide singularities, but an "antipodal" part of the universe.)

2 Analytic Continuation of Spacetimes

The purpose of spacetime, as originally conceived, is to describe the history of all inertial observers. A (timelike) geodesically incomplete spacetime fails to do this, so it behooves us to extend it as far as possible. (If even the maximal extension is incomplete we can begin to ask questions about cosmic censorship.) An analytic extension, if possible, is preferred because of its uniqueness and "permanence" (i.e., the continuation of Einstein's equations is automatic).

Actually, if not only the metric but also the manifold needs to be continued, then the continuation is not necessarily unique. A simple example is a finite part of (flat) Minkowski space, which can be continued either to the complete Minkowski space, or to one of the several locally Minkowskian spaces, such as the torus. More generally, any simply connected part of spacetime that can be continued to a multiply connected one can also be continued to a covering of the latter. We will encounter such ambiguities in the cosmological black hole case below. A somewhat more subtle example is the Taub-NUT space, which has two distinct and inequivalent analytic extensions [2]. In these ambiguous cases one needs to decide on such properties as the topology of the extended manifold along with the extension of the metric. Once the whole (smooth) manifold is known, its analytic structure is essentially unique. Because our manifolds are real, the relevant notion is real analyticity — existence of local power series expansions. Analyticity of the manifold means that the coordinate transformations between neighborhoods are real analytic functions.

In this context the analytic continuation of functions, such as the metric coefficients, is then unique, and the continuation satisfies the continued differential equation if the latter is itself analytic, as in the case of Einstein's equations.[1] To know this is, however, of little help in practical problems where the metric is given in some coordinates, and we desire to extend across a boundary where both the metric coefficients and the coordinates are non-analytic. There appears to be no systematic criterion for deciding whether analytic continuation is possible, and one usually has to rely on ingenuity to find suitable new coordinates in which the metric is analytic in the relevant region.

It can happen that neither the metric nor the coordinates are analytic functions on the manifold, but the metric coefficients are analytic functions of the coordinates; or they many be extendable to another (real)

---

[1] Analyticity is not assured, nor is it necessarily to be expected on physical grounds, if there are source terms present. Well-known examples are stellar models that are non-analytic on the stellar surface.

range of the coordinates by an excursion in the complex plane. Examples of this are found in the Schwarzschild metric at $r = 2m$ and $r = 0$ respectively. In either case we obtain solutions of the Einstein equations in the new coordinate range, but it is a separate question whether and how the geometry so described fits together with the original geometry, and if so, whether the fit is analytic. Finding a proper overlap seems to be the only way to assure the latter.[2] For a class of metrics to be discussed below, of which the Schwarzschild metric is a member, one knows how to fit together the pieces across horizons like $r = 2m$.

One can, of course, give criteria that establish *non*-analyticity, for example the divergence of invariants formed from the Riemann tensor and/or its derivatives. This happens, for example, at $r = 0$ in the Schwarzschild metric, so no real analytic extension is possible there. It is worth noting that, in the case of indefinite metrics, not all divergences of the Riemann tensor can be found in this way; this happens when there is a "null" infinity, as in a Riemann tensor of the type $R_{\mu\nu\alpha\beta} = l_{[\mu}m_{\nu]}l_{[\alpha}m_{\beta]}$, with $l^\mu l_\mu = 0$, $l^\mu m_\mu = 0$ and divergence in l, m. In that case the components in an orthonormal frame diverge. (To avoid spurious infinities due to the frame becoming null the orthonormal frame should be parallelly propagated).

If one admits an excursion to complex coordinate values one may find other real metrics analytically related to the original one. Such extensions are however not unique, and may have nothing directly to do with the original geometry.[3] In fact, the "extension" may have a different signature than the original metric. An example is the Euclidean Schwarzschild geometry that is used to describe instanton or thermal effects. An example of a Lorentzian, complex analytic but hardly physical relation of the

---

[2]It is remarkable that some solutions known in closed form (which is sometimes — loosely — called "analytic") can be extended with a high degree of differentiability (e.g. C^{122} as in [3]) but not analytically.

[3]To restore uniqueness it has been suggested [4] that the slightly complex path should be a geodesic. It remains to be seen whether one does not still lose physical significance in this unorthodox continuation.

Schwarzschild metric is

$$ds^2 = \left(1 - \frac{2m}{r}\right) dt^2 + \frac{dr^2}{\left(1 - \frac{2m}{r}\right)} - r^2 d\theta^2 + r^2 \cosh^2\theta \, d\phi^2.$$

Another example is the continuation of the de Sitter space metric,

$$ds^2 = -\frac{dt^2}{t^2} + e^{2Ht}(dx^2 + dy^2 + dz^2)$$

across $t = 0$, which effectively changes the cosmological expansion parameter H into its negative.

As remarked above, a null surface is a natural analyticity boundary, because "new" information can propagate along such surfaces. On the other hand, analyticity of a region would be expected to extend to the domain of dependence of that region. The features of simplicity that allow exact solutions may have a similar extent, so that the coordinates in which a metric is originally found tend to be analytic only in such domains, even when the geometry itself can be extended. Therefore an approach worth trying is to introduce null coordinates in which the boundary is one of the coordinate surfaces. The following class of metrics provides an example.

3   Walker's Spacetimes and Their Maximal Extension

Walker [5] considers spherically symmetric "static" metrics of the form

$$ds^2 = -Fdt^2 + \frac{dr^2}{F} + r^2 d\Omega^2$$

where $F = F(r)$ is the norm of the Killing vector $\partial/\partial t$. Here this Killing vector is not necessarily timelike (hence the quotes around "static") because we allow F to be positive or negative. F may be an analytic function of r, satisfying the Einstein equation, and range over positive and negative values, but the metric is clearly non-analytic at the zeros of F; the problem is to find the continuation across these zeros. (Infinities of

F imply infinities of the Riemann tensor, so no analytic continuation is possible there.) Because the angular part is regular for $r > 0$, it suffices to confine attention to the two-dimensional r, t part of the metric.

By "factoring" this two-dimensional part into two integrable null differential forms,

$$du = dt + \frac{dr}{F}, \qquad dv = dt - \frac{dr}{F} \qquad (1)$$

we can give the metric the double null form, $ds^2 = -F(u-v)\,dudv$, but as a metric this is still singular at $F = 0$. If instead, following Finkelstein's trick, we use r and only one null coordinate, say u, the metric assumes the nonsingular form

$$ds^2 = -F(r)du^2 + 2dudr,$$

which is analytic wherever F is analytic[4] as a function. This metric, then, provides the overlap necessary to connect two regions with opposite signs of F. This analytic connection between two neighboring regions is illustrated in the conformal diagram shown in Fig. 1. Here we have assumed that the region $r \to \infty$ has the usual asymptotically flat structure, and that there is another zero of F at finite r below $r = a$ (corresponding to the "roof" of the figure). If these structures are different, the blocks may not have a diamond shape, but the region around the zero, $r = a$, will look the same.

---

[4]In this sentence the word "analytic" could be validly replaced everywhere by "smooth" or "C^n". In those cases the extension would, of course, not be unique.

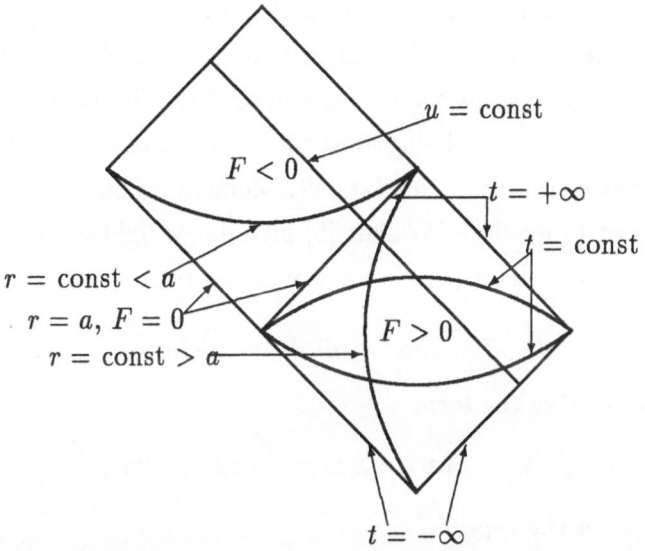

Fig. 1. Conformal diagram of region of Walker metric surrounding $r = a$, the largest zero of F. The r, t coordinates are degenerate on the boundaries of the diamond-shaped regions. For example, the lower left boundary can also be labeled $r = a$.

The spacetime as shown is still not complete. For example, the middle left boundary $r = a$ is at a finite distance, and so is the "roof." By using Finkelstein coordinates r and v we obtain a system that overlaps the middle left boundary, and by repeating this procedure around the next smaller zero below $r = a$ we can extend beyond the "roof." Thus in passing through each zero of F we add several new diamond-shaped regions to the conformal diagram, according as we cross the boundary along $u = $ const or $v = $ const. All the regions so generated at the zero $r = a$ are shown in Fig. 2.

The overlapping coordinates constructed so far reach across all of the diagonal lines, but not across the intersection points P, Q, R, ... These come in two types, those like P and R being characterized by the vanishing of the $\partial/\partial t$ Killing vector, and those like Q by different values of r trying to come together. It is therefore not surprising that no analytic continuation is possible or necessary across points of type Q; they are at an infinite distance ($t \to \infty$) along Killing orbits and not part of the manifold. (They can also provide a "safe haven" for observers who might otherwise experience a naked singularity.) At points of type P the blocks may fit together smoothly or analytically, depending on the form of F. The proof is not immediate; Walker [5] introduces lightlike coordinates U, V related to the u, v of (1) via an adjustable constant c,

$$dU/U = cdu \qquad dV/V = -cdv.$$

The metric then takes the form

$$ds^2 = G\,dU\,dV \qquad \text{with} \qquad G = \frac{F}{c^2} \exp\left(-2c \int \frac{dr}{F}\right).$$

For functions F of the type

$$F(r) = \frac{\prod_i (r - a_i)}{K(r)}, \tag{2}$$

with $K(r)$ a polynomial with zeros differing from the a_i, he shows that c can be chosen so that $G \neq 0$ at any one a_i.

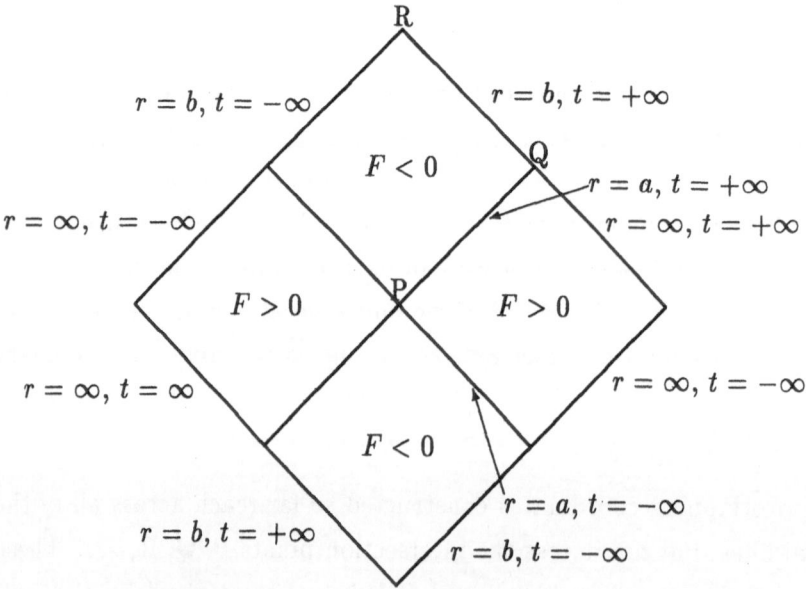

Fig. 2. Blocks that fit together at their $r = a$ boundary. The next lower zero of F occurs at $r = b$.

If two roots coincide, the picture looks different. There are no points of type P because the double-root horizon is at an infinite spatial distance; a spacelike section does not have the "wormhole" shape, but is an infinite funnel or "cornucopion." Figure 3 shows how the blocks fit together in that case [6]. All lines and curves that are shown correspond to $r =$ const. This diagram looks like what one would obtain by continuing Fig. 1 towards the upper left by the usual rules to another block with $F > 0$, and then eliminating the $F < 0$ block and moving the two $F > 0$ blocks together. Thus the coincidence limit of two roots of F does not appear continuous in the conformal picture. This happens, for example, for the Reissner-Nordström geometry, where the coincidence limit corresponds to an "extremally" charged black hole, $Q^2 \to M^2$. As long as the roots are distinct the two $F > 0$ blocks have a finite size $F < 0$ block between them. Instead of eliminating the $F < 0$ block, one can keep its physical size constant by rescaling the metric. The resulting spacetime is the Bertotti-Robinson universe. (More detail on the relation between the extremal Reissner-Nordström and the Bertotti-Robinson geometries is found in [7].)

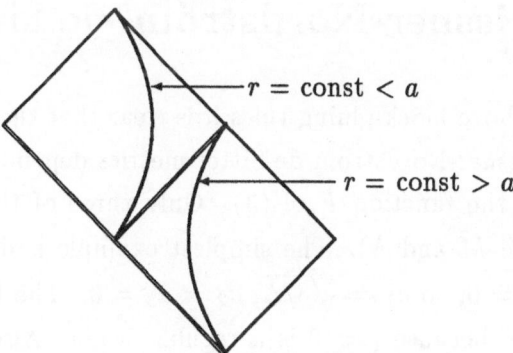

Fig. 3. Conformal diagram of region of Walker metric surrounding a double root of F, denoted by double lines.

Beyond the coincidence limit it can happen that a pair of (real) roots disappear. This is also not a continuous change in the conformal diagram — the two regions of Fig. 3 merge into one.

Metrics of the Walker type occur in the cosmological black hole context when there is a "single" black hole of mass M and charge Q in a universe with cosmological constant Λ. In that case, F takes the form

$$F(r) = \frac{(-\frac{1}{3}\Lambda r^4 + r^2 - 2Mr + Q^2)}{r^2} , \qquad (3)$$

which is of the type (2). (As there, we will denote the zeros of the numerator, in decreasing order, by $a_1 \ldots a_4$.) Thus we know that the maximal analytic extension is given by the Walker construction, and once we krow how the blocks fit together we can do all calculations in the original r, t coordinates.

4 Global Structure of de Sitter and Reissner-Nordström–de Sitter Cosmos

From the above block-gluing rules it is clear that the conformal diagrams for the Reissner-Nordström-de Sitter metrics depend only on the number of zeros of the function F of (3). Only three of the roots are positive (for positive M and Λ). The simplest example is de Sitter space itself, $M = 0$, $Q = 0$, so $a_1 = \sqrt{3/\Lambda}$, $a_2 = a_3 = 0$. The blocks look different in this case, because $r = 0$ is a regular origin. Also, $r = \infty$ is infinite distance in time (since $F < 0$ for $r > a_1$), so we can identify it with timelike and null infinity, $\Im$. Figure 4a shows an embedding of an r, t subspace of this geometry in flat 3-dimensional Minkowski space, and Fig. 4b is the corresponding conformal diagram. Note that the conformal diagram corresponds to only half of the embedded surface, because the latter shows both "sides" of the origin ($\phi = 0$ and $\phi = \pi$, for example).

De Sitter space is often described in coordinates different from the r, t used above, in which the space sections are conformally flat. These new coordinates r', t' are called *isotropic* or *cosmological* coordinates [8],

$$r = r'e^{Ht'} \qquad t = t' - \frac{1}{2H}\ln(1 - H^2r^2).$$

Here $H = \sqrt{\Lambda/3} = 1/a_1$ is the "Hubble constant" or cosmological expansion parameter. The metric then becomes

$$ds^2 = -dt'^2 + e^{2Ht'}\left(dr^2 + r^2 d\Omega^2\right). \qquad (4)$$

These coordinates cover more of de Sitter space than one patch of the r, t coordinates, but there still is a horizon at $t' = -\infty$. Another block of r', t' coordinates but with opposite sign of H can be analytically connected to the original one. We call the coordinates "expanding" in the block with $H > 0$, and "contracting" in the other one. Figure 5 shows some of the spacelike surfaces $t' = $ const.

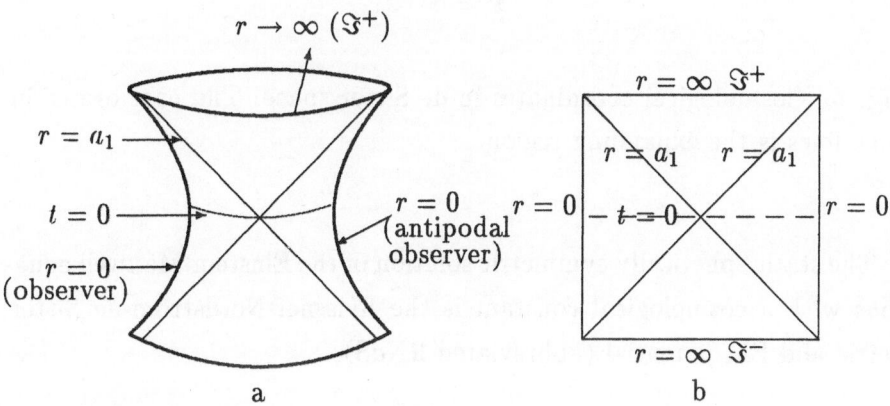

Fig. 4a. Embedding of 2-D de Sitter space in 3-D Minkowski space.

Fig. 4b. The corresponding conformal diagram.

That the de Sitter universe can appear as either static (3), or expanding or contracting (4), is an accident due to the high degree of symmetry of this model. In fact, each "observer" (timelike geodesic) has a static frame centered around him/her. When there is a "single" black hole present, a static frame still exists (but only the one that is centered about the black hole). The analytic extension of this case can therefore be easily treated by Walker's method. In discussing the global properties it is appropriate also to show the expanding and contracting, cosmological frames, because only their analog exists in the spacetimes [1] with several black holes.

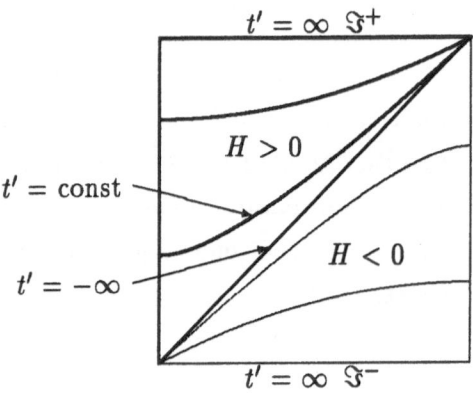

Fig. 5. Cosmological coordinates in de Sitter space. The part drawn in thick lines is the expanding region.

The static spherically symmetric solution of the Einstein-Maxwell equations with a cosmological constant is the Reissner-Nordström-de Sitter metric and EM potential (abbreviated RNdS),

$$ds^2 = -F\,dT^2 + \frac{dR^2}{F} + R^2\,d\Omega^2$$

$$F = 1 - \frac{2m}{R} + \frac{Q^2}{R^2} - \frac{1}{3}\Lambda R^2 \qquad A_T = -\frac{Q}{R}.$$

Here we have used capital letters to denote the static frame in order to distinguish it from the cosmological coordinates, which will be in lower case. By means of a somewhat involved coordinate transformation [8] one finds the form of the metric in cosmological coordinates,

$$ds^2 = -V^{-2}dt^2 + U^2 e^{2Ht}(dr^2 + r^2 d\Omega^2)\,, \tag{5}$$

where

$$U = 1 + \frac{M}{\rho} + \frac{M^2 - Q^2}{4\rho^2} \qquad V = \frac{U}{1 - \frac{M^2 - Q^2}{4\rho^2}} \qquad \rho = e^{Ht}r\,.$$

By means of the simple coordinate change

$$\tau \equiv H^{-1}e^{Ht}$$

we can extend the region covered by these coordinates, by allowing τ to be negative as well as positive. The metric and EM potential then become

$$ds^2 = -\frac{d\tau^2}{U^2} + U^2(dr^2 + r^2 d\Omega^2) \qquad U = H\tau + \frac{M}{r} \qquad A_\tau = \frac{1}{U}\,. \tag{6}$$

Figure 6 shows the analytic extension constructed according to the prescription of Sect. 3 for the generic case, when there are three roots of F.

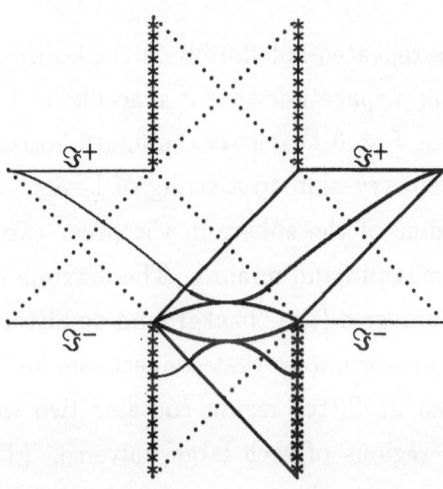

Fig. 6. Conformal diagram of the RNdS geometry. The diagonal (mostly dotted) lines are the horizons corresponding to the three roots of F that separate the different "static" blocks. Those crossing between $\Im^-$ and $\Im^+$

are the cosmological horizons; those crossing at the center of the figure are the outer black hole horizons; and those crossing at the top and at the bottom are the inner horizons. The multiply-crossed vertical lines are the singularities at $R = 0$. The thin curves describe surfaces of constant cosmological time τ, the upper one having $\tau > 0$, and the lower one $\tau < 0$. The cosmological r-coordinate is single-valued only to the right (or only to the left) of the point of contact of these curves with the lens-shaped region. This point of contact is a "wormhole throat" of the spacelike geometry. Surfaces of different τ-value touch the lens-shaped region at different points. The boundaries of the lens-shaped region are given by $R = \text{const} = M \pm \sqrt{M^2 - Q^2}$. The two region covered by the right-hand parts of these spacelike surfaces, in which the cosmological coordinates are therefore single-valued, is shown by the thick outlines. Note that the part near $\Im^+$ is similar to that shown in Fig. 5. The lens-shaped region in between is not covered by these coordinates. These regions as drawn are appropriate for expanding coordinates. A region covered by contracting coordinates is obtained, for example, by reflecting the regions in thick outline about the horizontal symmetry axis.

The figure can be repeated indefinitely in the horizontal and the vertical direction, yielding a spacetime that is spacelike and timelike periodic. The spacelike surface $T = 0$ that cuts the figure horizontally in its center then has the geometry akin to a string of beads: as we move along this surface, the radius of the sphere in the other two (θ, ϕ) directions alternately reaches maxima and minima. The maxima correspond to the large regions of the universe (the "background de Sitter space"), and the minima are throats of wormholes that connect one de Sitter region with the next. Thus each de Sitter region contains two wormhole mouths, placed in antipodal regions of each large universe. (This is the reason for the quotes above when calling this a "single" black hole in a de Sitter universe.) Alternatively and more compactly we can imagine the left and right halves of the figure identified, so that the horizontal spacelike surface of the figure is a closed circle, and the 3-D spacelike topology is

$S^1 \times S^2$. In this case the electric flux of the charge Q also describes closed circles: it emerges from one wormhole mouth, spreads out to the maximum universe size, reconverges on the other mouth, and flows through the wormhole back to the first mouth. Seen from the large universe, the first mouth appears positively charged, and the antipodal one, negatively charged — an example of Wheeler's "charge without charge"!

In this universe (as in the $\Lambda = 0$, asymptotically flat Reissner-Nordström (RN) geometry) a geodesic observer that wants to experience the singularity can do so, for example by moving along the vertical symmetry axis of the figure. In the RN case this is not considered a serious challenge to cosmic censorship, because the interior of the black hole, through which the observer in search of a singular experience must travel, is not stable under small perturbations of the exterior: radiation falling into the hole from the exterior would have a large blueshift at this observer — it would not only burn her up, but also change the nature of the singularity. It is remarkable that this does not necessarily happen in RNdS universes, for certain values of the parameters [9]. Thus these solutions are a counterexample to a strong interpretation of cosmic censorship. But there are, of course, many other geodesics that can lead observers who do not take the plunge to their safe haven at $\mathfrak{F}^+$.

The picture of Fig. 6 changes, for example as in Fig. 3, when roots of F coincide or cease to be real (see [8] and [10]). Another special case is of interest here because it can be generalized to the multi-black-hole case, namely $Q^2 = M^2$. In contrast to the RN case, when $\Lambda \neq 0$ this choice does not force a double root, so the global structure and conformal diagram of Fig. 6 still applies. What changes is the way the cosmological coordinates cover the diagram: the lens-shaped region degenerates into the line $\tau = 0$, so that a continuous region between the singularity and $\mathfrak{F}^+$ is covered. Therefore in this case the entire analytic extension of the metric (6) can be obtained by gluing together alternate copies of expanding and contracting cosmological coordinate patches. Figure 7 shows a pair of such regions, with the contracting one outlined by thick lines. (For easy comparison with Fig. 6 it may help to turn the latter upside down.) We see that the expanding patch contains $\Im^+$, and the contracting one contains $\Im^-$. It is therefore not immediately clear, if we calculate in contracting coordinates only, how to identify the black hole event horizon as the boundary of the past of $\Im^+$. We can follow outgoing lightrays to $r = \infty$, but at that point their geometrical distance R, measured by the Schwarzschild coordinate R (or by the area of the sphere $r = $ const) is still finite, $R = a_1$. However, there is this difference between such lightrays and those that fall into the black hole, that the latter reach the geometrical singularity, which *is* contained in the contracting cosmological coordinates. Furthermore, timelike geodesics heading toward the point S in the figure have an infinite proper time. Thus S is a safe haven for observers who desire to avoid the black hole, and it can be regarded as small piece of $\Im^+$ that can be asymptotically reached in contracting cosmological coordinates, just enough to be able to identify the event horizon.

The $Q^2 = M^2$ RNdS geometries still depend on two parameters, M and Λ resp. H, or on one dimensionless parameter $p = 4M|H|$ up to scale change. If $p < 1$ we have the "undermassive" case discussed so far. If $p > 1$ there is only one real root of $F(R) = 0$. The outer black hole horizon and the de Sitter horizon have disappeared, only what used to be the inner black hole horizon remains. One could also interpret the remaining horizon as a cosmological one, separating two naked singularities at antipodal regions of a background de Sitter space. The conformal diagram for this case, constructed according to the block gluing rules, is shown in Fig. 8.

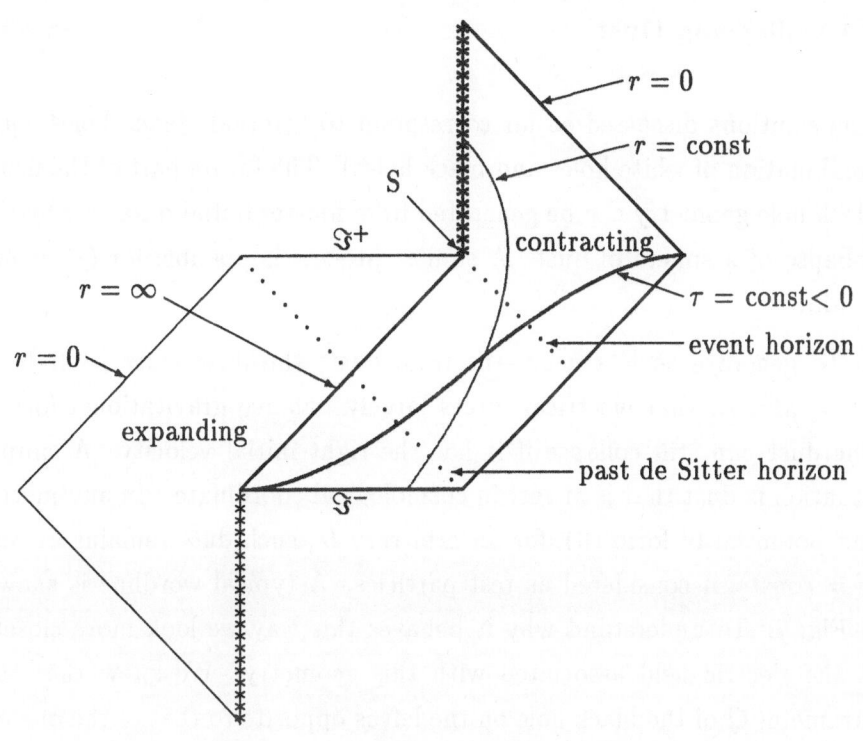

Fig. 7. Two patches of cosmological coordinates for the RNdS geometry for the case $Q^2 = M^2$ and $p < 1$ ("undermassive").

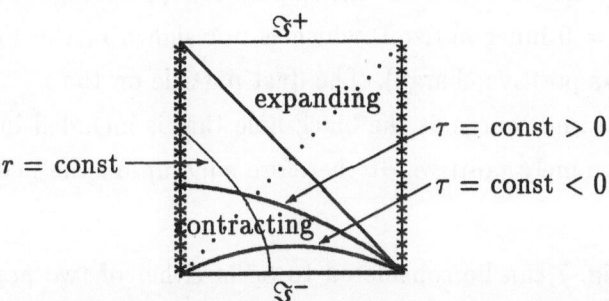

Fig. 8. Two patches of cosmological coordinates for the RNdS geometry for the case $Q^2 = M^2$ and $p > 1$ ("overmassive"). The thick outline shows the contracting patch.

4.1 Collapsing Dust

The solutions discussed so far correspond to "eternal" black holes (or a combination of white holes and black holes). The future part of the usual black hole geometry can be generated from matter initial data, say by the collapse of a sphere of dust. A similar process is possible for $Q^2 = M^2$ RNdS.

To generate such a geometry from dust, the dust must itself have $Q^2 = M^2$, so that electrical forces largely balance gravitational forces. The dust can still collapse if it has the right initial velocity. A simple situation is dust that is at rest in cosmological coordinates: in any metric and potential of form (6), for an arbitrary U, such dust remains at rest ($r = $ const), if considered as test particles. A typical wordline is shown in Fig. 7. To understand why it behaves this way we look more closely at the electric field associated with this geometry. We know that the parameter Q of the black hole on the left is opposite to that of the one on the right. Suppose the left hole is negative and the right one positive. On the spacelike surface corresponding to a horizontal line drawn through the center of Fig. 7, the electric field will then point from right to left. By flux conservation, the electric field will therefore also point to the left in the region near the upper singularity shown in the Figure. Thus we can associate a *negative* charge with this singularity. (The singularity to the right of the $r = 0$ inner horizon, which is not shown in the Figure, correspondingly has positive charge). The dust particle on the $r = $ const trajectory has the same charge as the black hole that is included in that coordinate patch, namely positive. It therefore ends up on the negative singularity.

Any point in Fig. 7 can be considered to be in either of two possible cosmological coordinate patches. For example, a point near $\Im^-$ is in the contracting patch shown, which contracts about the right black hole.

By reflecting this patch about a vertical line through the center of the Figure we obtain a patch that contracts about the left black hole. It also contains the region near $\Im^-$. Its radial coordinate will be denoted by r'. The trajectory $r' =$ const, obtained by reflecting the $r =$ const trajectory shown in the Figure, describes a negative test charge that falls into the positive singularity of the left (negative) black hole. Similarly the region between the de Sitter and the event horizon in the contracting patch has trajectories of positive charges that fall into the positive black hole, and of negative charges that go to $\Im^+$. The region inside the event horizon has trajectories that go to one or the other singularity, depending on their charges. Of course all these trajectories that are simply described by $r =$ const or $r' =$ const satisfy special initial conditions — their initial position is arbitrary, but their initial velocity is then determined.

To take into account the effect of the dust *on* the metric and potential, one needs to match the vacuum region to an interior solution. For spherical symmetry the matching conditions are equivalent to the demand that the dust at the boundary of the interior region move on a test particle path of the vacuum region. Thus a possible boundary for a collapsing (expanding) ball of dust is $r =$ const in collapsing (expanding) cosmological coordinates. The surface area $4\pi r^2 U^2$ of the dust ball collapses to zero when $U = 0$, which is also the location of the geometrical singularity; at that point the center of the dust ball must coincide with the surface. The corresponding conformal diagram therefore looks as shown by the thick curves in Fig. 9. The region filled with dots denotes the location of the dust.

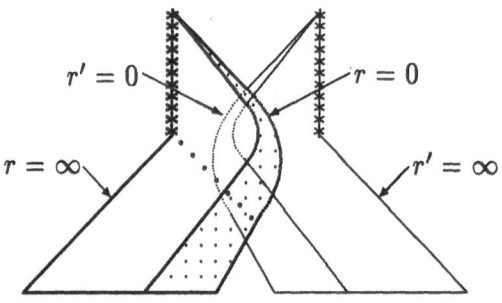

Fig. 9. Conformal diagram for a RNdS black hole generated by the collapse of a ball of charged dust in a de Sitter background (thick lines). The dust region is shown dotted. The dotted line is the black hole event horizon. The thin lines show a dust ball that is symmetrically placed at the antipodal region of the RNdS background, and has opposite charge. If both dust balls are present only the region between the curves $r = 0$ resp. $r' = 0$ applies.

Because the dust includes the origin $r = 0$, there is now no continuation to the right of the heavily outlined region necessary or possible. There is still a cosmological horizon, $r = \infty$, and the continuation on the other side could be analytic (a semi-infinite "string of beads") or reflection-symmetric about a vertical axis through $\Im^-$ (a collapsing dust ball at the antipodal region). Of the two geometrical singularities associated with one RNdS black hole, the dust covers up only the one that would have been on the right in the figure, with the same total charge as that of the dust. The other singularity could be covered up by another dust ball, with the opposite charge. This is shown by the thin curves in Figure 9. If both dust spheres are present, then the physical part of the geometry lies between the $r' = 0$ curve on the left and the $r = 0$ curve on the right. If conditions are as shown, they uncover and then cover up again a (small) region of vacuum RNdS geometry between them.

5 The Multi-Black-Hole Solutions

A solution representing any number n of arbitrarily placed charged black holes in a de Sitter background is given by a metric of type (6), with a different potential U [1]:

$$ds^2 = -\frac{d\tau^2}{U^2} + U^2(dr^2 + r^2 d\Omega^2) \qquad U = H\tau + \sum_{i=1}^{n} \frac{M_i}{|r - r_i|} \qquad A_\tau = \frac{1}{U}.$$

$$(7)$$

We will call this the KT solution. Each mass of the KT solution has a charge proportional to the mass, $Q_i = M_i$, and only the location r_i (not the initial velocity) is arbitrary. Here $|r - r_i|$ denotes the Euclidean distance between the field point r and the fixed location r_i in a Euclidean space of cosmological coordinates. For $n = 1$ this reduces to the $Q^2 = M^2$ case of the RNdS solution (6). Also, in the limit of large r, and for r_i in a compact region of Euclidean coordinate space, (7) approaches the RNdS solution with $M = \sum M_i$. One therefore expects the horizon structure at $r \to \infty$ to be similar to that of RNdS, which suggests that a $H > 0$ and a $H < 0$ version of (7) can be glued together as extensions of each other, similar to Fig. 7. The surprise is that, although this can be done with some degree of smoothness, it cannot be done analytically [11]. This means that there is no unique extension. An observer at rest in contracting ($H < 0$) cosmological coordinates, whose entire past can be described in these coordinates, has no way of telling what is in the other "half" of the de Sitter background (he can only guess that the total charge over there must be $-\sum M_i$, to balance the charge that he sees). If he moves and crosses the cosmological horizon $r = \infty$, the other, expanding half suddenly comes into view, seen at a very early time when all the masses are very close together. It is therefore reasonable that there is a pulse of gravitational and EM radiation associated with the horizon: this is the physical description of the lack of analyticity.

It is instructive to note how the various solutions differ in respect to possible coordinate choices. Pure de Sitter space has a static and two cosmological (expanding and contracting) coordinate systems centered about any timelike geodesic. In RNdS these coordinates are centered about the black hole only, but one can still choose between expanding and contracting frames near either of the holes. In KT there is one set of black holes, all with the same sign of charge, that is uniquely expanding (the distances between holes increasing as $e^{H\tau}$), and another, oppositely charged set that is uniquely contracting (the distances decreasing as $e^{-|H|t}$). The expanding set is described by cosmological coordinates that include $\Im^+$, whereas for the contracting set, $\Im^-$ is included.

5.1   Merging Black Holes

Since black (as opposed to white) holes are determined from $\Im^+$, it is easiest to identify the black hole horizons for the expanding set. Consider the expanding coordinates in RNdS as in the left half of Fig. 7. The boundary of the past of $\Im^+$ that lies in those coordinates is the left-hand line labeled $r = 0$ — it is the event horizon of the black hole in the "left" part of RNdS space. Similarly the event horizon of expanding KT space is given by $r - r_i = 0$. The Euclidean coordinate space with the n points r_i removed is a representation of $\Im^+$ of expanding KT space. The n missing points represent n disjoint boundary components, and there is a finite distance between them at all times. Thus we have n black holes that remain separate for all times.

It is not as easy to identify the black holes in the contracting KT space, because that does not contain much of $\Im^+$. But because the cosmological horizon at $r = \infty$ is so similar to that of the RNdS space it does contain a "point" like S of Fig. 7 that is just enough to define the event horizon. So, to find the event horizon while staying within one coordinate patch we must find the surface that divides lightrays reaching $r = \infty$ from those that reach the singularity. But where in the KT world is the singularity? One can show that metrics of type (7) are singular where $U = 0$. In the contracting case, $H < 0$ (and of course $M_i > 0$), this happens only for positive τ. Thus we need to find the "last" lightrays that just make it out to $r = \infty$ at $\tau = 0$. More precisely (since $r = \infty$ is not a very precise place) we need $H\tau$ finite in this limit.[5]

If we know these last lightrays (the 3D horizon surface) we can then intersect them with a spacelike surface to find the shape of the 2D horizon at different times. An interesting question is whether this 2D horizon changes topology with time. This would, for example, describe the merger of two black holes into one. If black holes merge we can try to make an overmassive (hence nakedly singular) one from two undermassive ones, to test cosmic censorship. This is of course possible only in a contracting part of the KT solution, because we have already seen that the black holes in the expanding part remain separate for all times.

All contracting black holes do eventually merge into one. To show this for the case of a pair of holes, we show that the horizon must consist of two parts at early times, and be a single surface at late times. One can see rather directly that light starting at r sufficiently close to any one of the r_i at any time and in any direction will reach the singularity, because it will spend its entire history in a geometry sufficiently close to the single black hole, RNdS geometry. Thus points sufficiently close to r_i will always lie within the event horizon. So, for the case of two black holes, a key question is what happens to light that starts on the midplane between the holes. In fact, if the light starts early enough it will always be able to escape to $r = \infty$. Thus at early times the midplane does not meet the horizon — the two black holes are disjoint. To show this we center our Euclidean coordinates at the midpoint between the holes, which have a Euclidean separation d. From $ds^2 = 0$ in (7) we then find, for radial outgoing null geodesics in the midplane,

$$\frac{d\tau}{dr} = U^2 = \left(H\tau + \frac{M}{\sqrt{r^2 + d^2}} \right)^2 < \left(H\tau + \frac{\sqrt{2}M}{r + d} \right)^2 .$$

---

[5]This is so because part of the "somewhat involved" coordinate transformation leading to (5) is actually rather simple, $R = Hr\tau + M$. For RNdS this is the static R, which is finite on the black hole horizon. Because near $r = \infty$ the KT geometry is so close to RNdS, R should also be finite for KT.

Now let

$$R_* = H\tau(r+d) \qquad y_* = \ln(r+d)$$

to find

$$(r+d)\,dR_*/dr > R_* + H\,(R_* + \sqrt{2}M)^2. \tag{8}$$

Standard analysis of this equation shows that if R_* is larger than the lower root of the RHS of (8), it will stay positive for all larger r. In terms of r and τ this means that if τ is sufficiently negative for any r (remember $H < 0$) then τ will remain negative as r increases — the lightray avoids the singularity. By a similar estimate one can show [11] that for each sufficiently late (but negative) τ there is a sphere surrounding both r_i such that all outgoing null geodesics will reach the $U = 0$ singularity. This means that the horizon surrounds both r_i, and the black holes have merged.

5.2   Continuing Beyond the Horizons

We could now discuss the merging of two undermassive into one over-massive black hole. Since the geometry near $r = \infty$ will be determined by the overmassive sum of the two individual masses, that neighborhood will look like the corresponding part of a single overmassive RNdS, that is, like the left hand side of Fig. 8. That contains a naked singularity (the left multiply-crossed line), which has nothing to do with the black hole merger, because it is located at the opposite side of the universe. But this is the only place in the coordinate patch where a $U = 0$ singularity occurs. Our patch does not describe enough of the history of the interesting region, where $|r - r_i|$ is small, just as the thick outline does not extend far to the right side of Fig. 8. To find out whether black hole merger generates its own naked singularity we must continue the KT metric beyond the inner black hole horizons at $r = r_i$. As we are continuing the KT geometry it is of course also interesting to look beyond the cosmological horizon at $r = \infty$, which exists only if the total mass is undermassive $(4|H|\sum M_i < 1)$. So we look at null geodesics that approach these horizons.

Let us choose the origin $r = 0$ of our Euclidean coordinates at the location of the ith mass (so that $r_i = 0$). The equation $ds^2 = 0$ satisfied by an ingoing null geodesic then takes the form, for small r,

$$\frac{d\tau}{dr} = -U^2 = -\left(H\tau + \frac{M}{r} + \sum_{j\neq i} \frac{M_j}{r_j}\right)^2. \tag{9}$$

We can eliminate the last (constant) term on the right by defining a new time coordinate $\tau' = \tau + H^{-1}\sum'(M_j/r_j)$. The equation then becomes an equality version of (8), and by analyzing it in the same way as above one finds [11] the limiting forms

$$2H^2 r\tau' \to 1 - 2M_i H - \sqrt{1 - 4M_i H}. \tag{10}$$

To assess any incompleteness we need to know how a null geodesic $(r(s), \tau(s))$ depends on the affine parameter s, and we can get that from the variational principle,

$$\delta \int \left(-\frac{1}{U^2}\left(\frac{d\tau}{ds}\right)^2 + U^2\left(\frac{dr}{ds}\right)^2\right) ds = 0.$$

The Euler-Lagrange equation for $\tau(s)$ together with the first equality of (9) yields

$$\frac{d^2 r}{ds^2} - 2HU\left(\frac{dr}{ds}\right)^2 = 0.$$

Substituting (10) we find, in the limit $r \to 0$,

$$\frac{d^2 r}{ds^2} - \frac{1 - \sqrt{1 - 4M_i H}}{r}\left(\frac{dr}{ds}\right)^2 = 0$$

with the solution

$$r \sim (s - s_{\text{hor}})^{\frac{1}{\sqrt{1-4M_i H}}}, \quad \text{hence} \quad \tau \sim (s - s_{\text{hor}})^{-\frac{1}{\sqrt{1-4M_i H}}}. \tag{11}$$

So the inner horizon is reached at a finite parameter value s_{hor}. Similarly one finds that the cosmological horizon is reached in a finite parameter interval,

$$r \sim (s - s_{\text{Hor}})^{-\frac{1}{\sqrt{1+4(\Sigma M_i)H}}}, \quad \tau \sim (s - s_{\text{Hor}})^{\frac{1}{\sqrt{1+4(\Sigma M_i)H}}}. \tag{12}$$

This behavior of the coordinates r and τ gives us important information about the differentiability and analyticity of the geometry near the horizon. We can first eliminate whichever of the two is infinite on a given horizon in favor of $\hat{R} = Hr\tau$, which is always finite on the horizon. The metric is then an analytic function of remaining, finite coordinates, so the Riemann tensor will also be analytic in these coordinates. But in order that the geometry be differentiable, the Riemann tensor should be differentiable in the affine parameter s along null geodesics. Thus the differentiability of the geometry is measured by that of r resp. τ as a function of s, as given by (11) and (12).

Consider first the neighborhood of the inner horizon, where r is finite. Since $H < 0$ we have $1/\sqrt{1 - 4M_i H} < 1$, r is not a differentiable function of s at $r = 0$, and the Riemann tensor will be singular there. A more careful analysis [11], using a transformation to coordinates that are not singular on the horizon, shows that the metric is C^1 but in general not C^2. There is therefore no unique, analytic extension across the inner horizon. One can match differentiably essentially any KT solution with the same mass M_i. One can increase the differentiability by arranging the other masses carefully around the ith one, so as to make the potential U approximately spherically symmetric (by eliminating multipoles to some order). The neighborhood of M_i then becomes approximately RNdS and hence "more nearly analytic" — i.e., of increased differentiability.

The situation near the cosmological horizon offers more variety. To have this horizon at all the total mass must be undermassive, $4|H|\Sigma(M_i) < 1$. Here τ is the finite one, and the corresponding power of s is $1/\sqrt{1 - 4(\Sigma M_i)|H|} > 1$. Thus τ is always at least C^1. The transformation to coordinates that are good on the horizon shows that the metric is always at least C^2. In the special cases when the power is an integer n, i.e., for masses such that

$$4H \sum M_i = 1 - \frac{1}{n^2},$$

the metric is C^∞. For these values the smooth continuation matches the KT spacetime at the cosmological horizon to one with the same position and magnitudes of all the masses (so that all multipole moments agree), but with the opposite sign of H. We do not understand the physical significance of these special masses.

To show that the geometry at these horizons is not more differentiable than claimed one can compute the Riemann tensor. This infinity is of the null type mentioned in Sect. 2, and does not show up in invariants formed from the Riemann tensor. To see this for the horizon at $r = 0$ (or $r = \infty$) we write the metric (7) in terms of the coordinate $\hat{R} = Hr\tau$ and $y = \ln r$, and the quantity $W = rU$. Then the horizon occurs at $r \to \pm\infty$, where $\hat{R}$ and W are finite,

$$ds^2 = -\frac{(d\hat{R} - \hat{R}dy)^2}{H^2 W^2} + W^2(dy^2 + d\Omega^2).$$

Now an invariant formed from the curvature tensor involves terms in derivatives of the metric and its inverse, multiplied by powers of the metric and its inverse. All these reduce to derivatives of W and $\hat{R}$ divided by powers of W. But all derivatives of W remain bounded as $y \to \pm\infty$, and W is finite on the horizon. Thus the invariants cannot blow up.

Singularities do show up in the components of the Riemann tensor in a parallelly propagated frame, for example along the null geodesic $(r(s), \tau(s))$ discussed above. Let $l = \partial/\partial s$ be the parallelly propagated tangent. Because of the asymptotic symmetry near the horizon, $\eta = \partial/\partial\theta$ is also asymptotically parallelly propagated. Now the frame component $R_{\mu\nu\rho\sigma} l^\mu \eta^\nu l^\rho \eta^\sigma$ contains the term $g_{\theta\theta,ss}$. If $g_{\theta\theta} = W^2$ depends on r (and not just on the regular R), it will not be a smooth function of s. In the RNdS ("single mass") case, $g_{\theta\theta}$ depends only on R. In the KT case the corrections to that behavior near an inner ($r = 0$) horizon start with the power $r^2 = (s - s_{\text{hor}})^{\frac{2}{\sqrt{1-4M_iH}}}$ unless there is special symmetry; thus one finds the differentiability of the metric as claimed above.

6 Naked Singularities?

Naked singularities visible to observers safely outside the strong curvature regions do not form in realistic gravitational collapse — this is the essential notion behind cosmic censorship. It can be made more precise in various ways [12]. At present none of these have been proved to be true in generic cases. When a proof seems difficult, it may be easier to obtain a convincing counterexample. Even if the conjecture is correct under certain assumptions, counterexamples are useful to test the necessity of these assumptions. For example, it may be that a version of cosmic censorship holds in pure general relativity, but fails when the theory is modified, say by a cosmological constant or by applying it to higher dimensions. There are in fact indications that cosmic censorship fails in the higher-dimensional theory inspired by string theory: certain 5D black strings (objects that appear four-dimensionally like black holes) are unstable, it is entropically favorable for them to decay into a set of 5D black holes, and during this decay naked singularities would form [13]. In the present contribution we want to test whether cosmic censorship fails in the special circumstances that are afforded when a cosmological constant is present.[6]

The idea of using KT spacetimes to test cosmic censorship is to start with two or more small (and hence not naked) black holes and let them collapse to form a single large (and maybe naked) one. We have seen that the KT solutions indeed can describe coalescing black holes, in the sense that the event horizons coalesce. But if it is possible to define the event horizon as we did, by observers who live for an infinite proper time in one KT coordinate patch, then these observers will see no signal from any singularity — all singularities that form from the initial data, including those in any of the (non-analytic) extensions, lie inside this event horizon. To find a situation where there is no "safe haven," so that the generic observers does see a singularity, we must suppose that $\sum M_i$ is overmassive, so the initial black holes cannot be defined by their

---

[6]A similar test for Einstein-Maxwell-dilaton theory with a cosmological constant inspired by string theory has been discussed in [14]

event horizon. An alternative to starting with black holes that have event horizons is to start with regular initial data on a compact surface. The KT solution cannot provide this either, because each M_i has an infinite throat. But such throats are the next best thing: each throat is undermassive, it is surrounded by a trapped surface, so one would not expect that the asymptotic regions down the throat could influence the solution in the interior. Can we, then, construct a KT solution of undermassive throats that has regular initial data and a naked singularity in its time development?

To decide this we must explore the global structure of the KT geometry. Unfortunately this cannot be completely represented by 2D conformal diagrams, because there is insufficient symmetry to suppress the additional dimensions. If we confine attention to the case of two equal masses, they lie on a line in the Euclidean coordinate space (which is an axis of symmetry of the spacetime). We can represent the essential features of the spacetime by drawing the conformal diagram for the spacetime spanned by the part of the axis going from one of the masses to infinity. This is shown in Fig. 10a. The part of the diagram representing the region near $r = \infty$ is to be read like a normal conformal diagram (i.e., each point represents a 2-sphere), whereas the region near $r = 0$ is to be thought of as doubled (each point represents two 2-spheres).

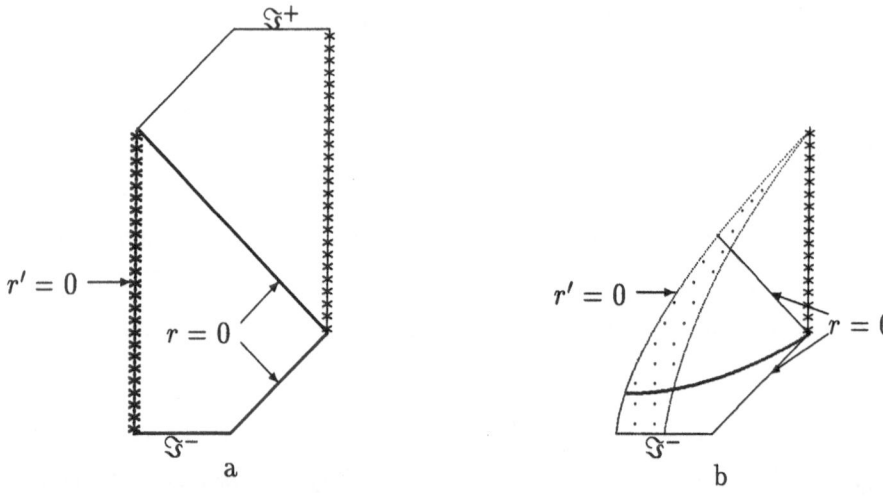

Fig. 10a. Conformal diagram of history of axis from one black hole to "infinity" for a two-black-hole contracting KT geometry. The black holes correspond to the region on the right, and the surrounding "de Sitter background" is the region on the left. The total mass is assumed to be overmassive, so this "background" is really an overmassive RNdS geometry with the singularity shown on the left. Therefore the axis extends to $r = \infty$ only for $\tau < 0$, after that it hits this singularity at $U(r, \tau) = 0$. The C^1 extension across the upper $r = 0$ horizon was chosen to be the time-reverse of the heavily outlined region.

Fig. 10b. In this diagram the cosmological singularity of Fig. 10a is covered up by a sphere of dust, as discussed in Sect. 4.1. The dust region is shown dotted. The thick curve is a spacelike surface with nonsingular initial data, containing two infinite charged black hole throats and, on the antipodal point of the universe, a collapsing sphere of dust. All observers have to cross the future $r = 0$ Cauchy horizon and can thereafter see the naked singularity shown on the right, the result of the merger of the two black holes. The spacetime ends after a finite proper time in a "big crunch."

The region near $r = \infty$ of a KT solution (7) always behaves like an RNdS geometry with mass equal to the total mass $\sum M_i$. If this is overmassive, there will be a curvature singularity in this region, whether the individual (undermassive) holes have merged or not. This singularity at the antipodal point of the universe has nothing directly to do with the black hole merger. To obtain nonsingular initial data we can eliminate this singularity by replacing it with a collapsing sphere of dust, as in Sect. 4.1. The resulting diagram is shown in Fig. 10b. The initial data induced on the spacelike surface shown by the thick curve are now nonsingular. In the time development shown, beyond the future Cauchy horizons $r = 0$, a curvature singularity appears. It comes in from infinity through the infinite throats of the merged black holes and spreads to $r' = 0$, i.e. to the antipodal point of the universe. The fact that the infinite throats are hidden behind trapped surfaces does not seem to be sufficient to prevent the singularity from coming "out of" the throat. Perhaps a less coordinate-oriented way of saying this is that all of space collapses down the throats, carrying all observers with it.

It is clear from the diagram that all observers originating on the initial surface will reach the Cauchy horizon, and if they extend beyond, they will see the singularity. We have seen that the Cauchy horizon surrounding a typical KT throat is not smooth, so that delicate observers may not survive the crossing. But by distributing several KT masses symmetrically about a given one, we can make that one as differentiable as necessary to ensure an observer's survival. So it is reasonable to conclude that cosmic censorship is violated in these examples.

The initial data in these examples already contain the black holes' infinite throats, and are not compact. Can we first form the black holes from collapsing dust, and then let them go through the above scenario? We have seen in Fig. 9 that we can have simultaneous collapse of a dust ball to form a black hole, and simultaneously remove the overmassive singularity at the antipodal point by another dust ball. The problem is now that in the KT solution, as in the RNdS case shown in Fig. 9, the two balls collide before any singularities have formed, at least if the balls move on $r = 0$ resp. $r' = 0$ trajectories in KT (cosmological) coordinates. Even if we allow more general trajectories we know for charged test particles that the trajectories tend to avoid singularities of the same charge; and even if naked singularities were formed later in the evolution, one would not know whether they were a fundamental property of the theory, or due to the dust approximation ("shell crossing singularities," which occur also in the absence of gravity and hence have nothing to do with cosmic censorship).

Even if we accept the KT solution's infinite throats in place of compact initial data, we still do not yet have a serious violation of cosmic censorship, because the general KT solution is still quite special. The initial position and masses can be specified arbitrarily, but not the initial velocities. The constraints on the initial values can be solved in more general (but still not quite generic) contexts, for example one can drop the $Q_i^2 = M_i^2$ condition [11]. These initial data can be analytic, but we do not know what happens beyond the Cauchy horizon. In the general KT solution we have seen that one has to cross the Cauchy horizon to see the naked singularity. It is not clear whether more generic solutions have a Cauchy horizon with a stronger singularity than the KT solution. If so, then cosmic censorship would be preserved.

References

[1] D. Kastor and J. Traschen, *Phys. Rev. D* **47** (1993) 5370.

[2] P. Cruściel and J. Isenberg, *Phys. Rev. D* **48** (1993) 1616.

[3] P. Cruściel and D. Singleton, *Commun. Math. Phys.* **147** 137 (1992)

[4] K. Peeters, C. Schweigert and J. van Holten, "Extended Geometry of Black Holes", preprint gr-qc/9407006.

[5] M. Walker, *J. Math. Phys.* **11** (1970) 2280.

[6] K. Lake, *Phys. Rev. D* **20** (1979) 320.

[7] B. Carter in *Black Holes* ed. C. DeWitt and B.DeWitt (Gordon & Breach 1973); D. Brill, *Phys. Rev. D* **46** (1992) 1560.

[8] D. Brill and S. Hayward, *Class. Quantum. Grav.* **11** (1994) 359.

[9] F. Mellor and I. Moss, *Phys. Rev. D* **41** (1990) 403 and *Class. Quantum Grav.* **9** (1992) L43; also see P. Brady and E. Poisson, *Class. Quantum Grav.* **9** (1992) 121; P. Brady, Núñez and Sinha, *Phys. Rev. D* **47** (1993) 4239; C. Chambers and I. Moss, *Class. Quantum Grav.* **11** (1994) 1035.

[10] K. Lake, *Phys. Rev. D* **19** (1979) 421.

[11] D. Brill, G. T. Horowitz, D. Kastor and J. Traschen, *Phys. Rev. D* **49** (1994) 840.

[12] See, for example, V. Moncrief and D. Eardley, *Gen. Rel. Grav.* **13** (1981) 887; R. Wald *General Relativity*, University of Chicago Press 1984; P. Joshi, *Global Aspects in Gravitation and Cosmology*, Oxford 1993 and the references cited therein.

[13] R. Gregory and R. Laflamme, *Phys. Rev. Lett.* **70** (1993) 2837.

[14] J. Horne and G. Horowitz, *Phys. Rev. D* **48** (1993) R5457.

The Structure of Quantum Conformal Superspace

Arthur E. Fischer[1] and Vincent Moncrief[2]

[1] Department of Mathematics, University of California, Santa Cruz, California 95064. Email: aef@cats.ucsc.edu

[2] Departments of Mathematics and Physics, Yale University, New Haven, Connecticut 06511. Email: moncrief@yalph2.physics.yale.edu

Abstract. For a compact connected orientable n-manifold M, $n \geq 3$, we study the structure of *classical superspace* $\mathcal{S} \equiv \frac{M}{\mathcal{D}}$, *quantum superspace* $\mathcal{S}_0 \equiv \frac{M}{\mathcal{D}_0}$, *classical conformal superspace* $\mathcal{C} \equiv \frac{M/\mathcal{P}}{\mathcal{D}}$, and *quantum conformal superspace* $\mathcal{C}_0 \equiv \frac{M/\mathcal{P}}{\mathcal{D}_0}$. The study of the structure of these spaces is motivated by questions involving reduction of the canonical Hamiltonian formulation of general relativity to a non-degenerate Hamiltonian formulation, and to questions involving both linearization stability and quantization of the gravitational field. We show that if the degree of symmetry of M is zero, then $\mathcal{S}, \mathcal{S}_0, \mathcal{C}$, and $\mathcal{C}_0$ are ILH-*orbifolds*. The case of most importance for general relativity is dimension $n = 3$. In this case, for a broad class of 3-manifolds for which $\deg M = 0$, we show that quantum superspace $\mathcal{S}_0$ and quantum conformal superspace $\mathcal{C}_0$ are in fact ILH-*manifolds*. If M is a Haken 3-manifold with $\deg M = 0$, then quantum superspace and quantum conformal superspace are *contractible* ILH-manifolds. Under these circumstances, there are no Gribov ambiguities for the configuration spaces $\mathcal{S}_0$ and $\mathcal{C}_0$. Our results are also applicable to the problem of reduction of Einstein's vacuum equations, to linearization stability, and to the problem of quantization of the gravitational field. Our results can be used to reduce the canonical Hamiltonian formulation, together with its constraint equations, to an unconstrained

Hamiltonian system. On the reduced phase space the canonical variables are free, or unconstrained, and carry complete information about the true degrees of freedom of the gravitational field. The structure of the reduced phase space is also of importance in understanding certain questions involving linearization stability and quantization of the gravitational field. For questions regarding linearization stability, the cotangent bundle of $\mathcal{C}$ plays a role in understanding the symplectic structure of the space of true degrees of freedom of the gravitational field. For questions regarding quantization, the space $\mathcal{C}_0$ plays the role of the reduced configuration space for quantum gravity.

1 Introduction, notation and background

The problem that we are concerned with in this paper is the structure of superspace and conformal superspace when the underlying manifold M is compact and has degree of symmetry zero. Our results are applicable to the reduction of Einstein's vacuum equations, to linearization stability, and to the quantization of the gravitational field (see Moncrief ([1989], [1990]) and Fischer-Moncrief ([1994a,b], [1995a,b]) and the references cited therein). The problem of reduction involves writing Einstein's vacuum equations as an *unconstrained* dynamical system where the variables are the *true degrees of freedom* of the gravitational field. In the reduced system, the constraint equations do not appear as they have already been eliminated from the theory. For applications involving linearization stability, we are able to find sufficient topological conditions on a spatially compact constant mean curvature Cauchy hypersurface that ensure that the resulting globally hyperbolic Ricci-flat spacetime is linearization stable. For applications relating to the quantization of the gravitational field, we propose applying our reduced Hamiltonian to a quantization procedure in which the constraints are eliminated before quantizing. In this approach, quantization would proceed using quantum conformal superspace as the reduced configuration space for quantum gravity. We will discuss these applications in detail elsewhere. Here we are interested in the structure of the configuration spaces that arise in these theories.

Let M be a compact connected smooth (C^∞) n-manifold, $\mathcal{M} = \text{Riem}(M)$ the space of smooth Riemannian metrics on M, $\mathcal{D} = \text{Diff}(M)$ the group of smooth diffeomorpisms of M, $\mathcal{D}_0$ its connected component of the identity, $\mathcal{P} = \text{Pos}(M)$ the abelian group (under pointwise multiplication) of smooth positive functions on M, and $\mathcal{D} \times \mathcal{P}$ the semi-direct product of $\mathcal{D}$ and $\mathcal{P}$. *Classical superspace* (or just *superspace*) is defined as the space of Riemannian geometries

$$\mathcal{S} \equiv \mathcal{M}/\mathcal{D},$$

pointwise conformal superspace is defined as the space of *pointwise conformal structures*

$$\mathcal{M}/\mathcal{P},$$

and *classical conformal superspace* (or just *conformal superspace*) is defined as the space of conformal geometries

$$\mathcal{C} \equiv \frac{\mathcal{M}/\mathcal{P}}{\mathcal{D}} \approx \frac{\mathcal{M}}{\mathcal{D} \times \mathcal{P}}.$$

We are also interested in the $\mathcal{D}_0$-restricted counterparts of superspace and conformal superspace; namely, $\mathcal{D}_0$-*restricted superspace*, or *quantum superspace*,

$$\mathcal{S}_0 \equiv \mathcal{M}/\mathcal{D}_0$$

and $\mathcal{D}_0$-*restricted conformal superspace*, or *quantum conformal superspace*,

$$\mathcal{C}_0 \equiv \frac{\mathcal{M}/\mathcal{P}}{\mathcal{D}_0} \approx \frac{\mathcal{M}}{\mathcal{D}_0 \times \mathcal{P}}.$$

This terminology is motivated by arguments that the appropriate configuration space for quantum gravity is $\mathcal{S}_0$ (see e.g. Friedman and Witt [1986], Witt [1986], Balachandran [1989], Sorkin [1989], Friedman [1990], and the references cited therein). These arguments are based on the fact that in quantum gravity a complex-valued wave function $\Psi : \mathcal{M} \to C$ satisfies the quantum momentum constraint

$$\delta_g(\frac{\hbar}{i}\frac{\delta\Psi}{\delta g}) = 0$$

if and only if Ψ is $\mathcal{D}_0$-invariant. Here $\frac{\delta\Psi}{\delta g}$ denotes the functional derivative of Ψ with respect to g and δ_g denotes the covariant divergence with respect to g; in local coordinates, $(\delta_g(\frac{\hbar}{i}\frac{\delta\Psi}{\delta g}))^i = -(\frac{\hbar}{i}\frac{\delta\Psi}{\delta g_{ij}})_{|j} = 0$. The main point is that a wave function Ψ on $\mathcal{M}$ need only be $\mathcal{D}_0$-invariant and not fully $\mathcal{D}$-invariant in order to satisfy the quantum momentum constraint. From this fact it is then argued that a (partially) reduced configuration space for quantum theory of gravity should be $\mathcal{S}_0$ rather than $\mathcal{S}$. These arguments may be extended to conformal superspace as well to argue that the reduced configuration space for a quantum theory of gravity that is based on a reduced Hamiltonian formulation should be $\mathcal{C}_0$ rather than $\mathcal{C}$. Although not definitive, these arguments add weight to the important role played by $\mathcal{S}_0$ and $\mathcal{C}_0$ in the study of quantum gravity; see also Section 6 for other arguments for why the study of quantum superspace and quantum conformal superspace are important.

We also remark that for orientable 2-manifolds, $\mathcal{C}_0$ is precisely the Teichmüller space of the manifold (see Earle-Eells [1969], Fischer-Tromba ([1984a,b,c], [1987]), and Tromba [1992] for more information regarding this viewpoint, and also Theorem 8.4). Thus $\mathcal{C}_0$ may also be thought of as a generalization of Teichmüller space to n-dimensions, $n \geq 3$ (see Theorems 8.2 and 8.3).

As has been emphasized in previous studies of superspace (Fischer [1970]), the structure of superspace and conformal superspace is linked to the topology of M. This linkage has resulted in the Stratification Theorem for classical superspace. Using the results of Fischer-Marsden ([1977]), a similar stratification theorem can be proven for conformal superspace. In this paper, we are interested in other aspects of this linkage; namely, we are interested in finding topological conditions on M that force the various orbit spaces $\mathcal{S}, \mathcal{S}_0, \mathcal{C}$, and $\mathcal{C}_0$ to have certain additional structures. These additional structures increase as we consider more topological conditions on the underlying manifold M. Thus, with increasing topological restrictions on M, we first get ILH-orbifold structures on $\mathcal{S}, \mathcal{S}_0, \mathcal{C}$, and $\mathcal{C}_0$, then ILH-manifold structures on $\mathcal{S}_0$ and $\mathcal{C}_0$, and finally contractible ILH-manifold structures on $\mathcal{S}_0$ and $\mathcal{C}_0$. Here ILH denotes Inverse Limit Hilbert (see Section 2 and Definition 6.2); in particular, these spaces will be infinite-dimensional orbifolds or manifolds. We will link the structure of these spaces to the underlying topology of M.

Anticipating our terminology, we summarize our results as follows (see Theorems 6.3 and 7.1):

Theorem 1.1 (Orbifold structure theorem) *Let M be a compact connected orientable n-manifold, $n \geq 3$, and $\deg M = 0$. Then superspace $\mathcal{S} = \mathcal{M}/\mathcal{D}$, quantum superspace $\mathcal{S}_0 = \mathcal{M}/\mathcal{D}_0$, conformal superspace $\mathcal{C} = \frac{\mathcal{M}/\mathcal{P}}{\mathcal{D}}$, and quantum conformal superspace $\mathcal{C}_0 = \frac{\mathcal{M}/\mathcal{P}}{\mathcal{D}_0}$ are connected smooth ILH-orbifolds.*

The case $n = 3$ is of the most importance for applications to general relativity. In this regard, we have the following (see Theorem 2.11, Eqn 2–2, and Theorem 8.2):

Theorem 1.2 (Manifold structure theorem) *Let M be a compact connected orientable 3-manifold. Assume that either M is an irreducible $K(\pi, 1)$-manifold with $\deg M = 0$, or that M is a com-*

posite manifold (i.e., a connected sum of two or more primes) of the form

$$M \approx \underbrace{S^3/\Gamma_1 \# \ldots \# S^3/\Gamma_k}_{spherical\text{-}factors} \# \underbrace{(S^1 \times S^2)_1 \# \ldots \# (S^1 \times S^2)_l}_{handles\ (or\ wormholes)}$$

$$\# \underbrace{K(\pi_1, 1) \# \ldots \# K(\pi_m, 1)}_{K(\pi,1)\text{-}factors}$$

where either M has at least one $K(\pi, 1)$-factor in its prime decomposition, or if it does not, then it has at least one spherical-factor S^3/Γ_i in its prime decomposition that is not a lens space.

Then quantum superspace $S_0 = \mathcal{M}/\mathcal{D}_0$ and quantum conformal superspace $\mathcal{C}_0 = \frac{\mathcal{M}/\mathcal{P}}{\mathcal{D}_0}$ are connected and simply connected ILH-*ma ni folds.*

We remark that the $\deg M = 0$ condition gives 3-manifolds that are not the familiar manifolds that relativists are used to dealing with. On the other hand, this class of manifolds is generic inasmuch as "most" 3-manifolds satisfy this condition (see e.g. Ebin [1970]).

Increasing the topological restrictions on M then gives the following stronger result (see Section 5 for definitions and Theorem 8.3):

Theorem 1.3 (Contractible manifold structure theorem) *Let M be a compact connected orientable Haken 3-manifold with $\deg M = 0$. Then quantum superspace S_0 and quantum conformal superspace $\mathcal{C}_0$ are contractible* ILH-*manifolds.*

Acknowledgments: We would like to thank Professors Domenico Giulini, Geoffrey Mess, and Frank Raymond for helpful conversations. We would also like to acknowledge the gracious hospitality and financial support of the *Erwin Schrödinger Institute for Mathematical Physics* in Vienna, where part of this research was carried out, and to thank the conference organizers Peter Aichelberg and Robert Beig for cordially inviting us to the Institute during the Summer of 1994 for the *Conference on Mathematical Relativity*, July 25-29, 1994, and for the *Workshop on Mathematical Relativity* following the Conference. We would also like to thank Spiros Cotsakis and Gary Gibbons, the organizers of the *First Samos Meeting on Cosmology, Geometry, and Relativity*, August 5-7, 1994, and the *Mathematics Department of the University of the Aegean*, Samos, Greece, for their kind hospitality and financial support. Vincent Moncrief would like to acknowledge support for this research by NSF Grants PHY-9201196 and INT-9015153 to Yale University.

2 Some topological and geometrical preliminaries

In this section we present some topological and geometrical background information that we shall need. Throughout this paper, an *n-manifold M* will mean a smooth (C^∞) Hausdorff second countable *n*-dimensional manifold without boundary. All tensor fields, including all Riemannian metrics, all maps between manifolds, and all group actions will be smooth.

For M a connected n-manifold, let

$\mathcal{M} = \text{Riem}(M) =$ the space of smooth Riemannian metrics on M

$\mathcal{D} = \text{Diff}(M) =$ the group of smooth diffeomorphisms of M

$\mathcal{D}_0 = \text{Diff}_0(M) =$ the connected component of the identity of $\mathcal{D}$

$\mathcal{X} =$ the space of smooth vector fields on M

$\mathcal{F} = C^\infty(M, \mathbf{R}) =$ the space of smooth real-valued functions on M

$\mathcal{P} = \text{Pos}(M) =$ the space of smooth real-valued positive functions on M

If M is orientable, let $\mathcal{D}^+ = \mathrm{Diff}^+(M)$ be the group of orientation–preserving diffeomorphisms of M; in this case, $\mathcal{D}_0 \subseteq \mathcal{D}^+$. If M is compact, each of these spaces can be given the structure of a smooth infinite dimensional ILH-manifold (Inverse Limit Hilbert); see Omori [1970], Ebin [1970], Ebin-Marsden [1970], or Fischer-Marsden [1975] for more details regarding these spaces.

For a Riemannian metric $g \in \mathcal{M}$, (M, g) denotes the corresponding Riemannian manifold. For $g \in \mathcal{M}$, let

$$I_g = I_g(M) = \mathrm{Isom}_g(M) = \{f \in \mathcal{D} \mid f^*g = g\}$$

denote the group of isometries of g, let $I_g^0 = I_g^0(M)$ denote its connected component of the identity, and let

$$\mathcal{I}_g = \mathcal{I}_g(M) = \{X \in \mathcal{X} \mid L_X g = 0\}$$

denote the Lie algebra of Killing vector fields of (M, g). Here $L_X g$ denotes the Lie derivative of g with respect to the vector field X.

Similarly, let

$$C_g = C_g(M) = \mathrm{Conf}_g(M) = \{f \in \mathcal{D} \mid f^*g = pg \text{ for some } p \in \mathcal{P}\}$$

denote the conformal group of g, let $C_g^0 = C_g^0(M)$ denote its connected component of the identity, and let

$$\mathcal{C}_g(M) = \{X \in \mathcal{X} \mid L_X g = \sigma g \text{ for some } \sigma \in \mathcal{F}\}$$

denote the Lie algebra of conformal Killing vector fields of (M, g).

A useful concept for a connected n-manifold M is its degree of symmetry (Hsiang [1967a], [1967b], and [1971]):

Definition 2.1 *For M a connected n-manifold, the* **degree of symmetry** *of M is defined by the non-negative integer*

$$\deg M = \max\{\dim I_g(M) \mid g \in \mathcal{M}\} \le \tfrac{1}{2}n(n+1)$$

Note that deg M is independent of any Riemannian metric on M and thus is a differential-topological invariant, rather than a geometrical invariant.

From the definition of degree, deg $M = 0$ is equivalent to any of the following: For each $g \in \mathcal{M}$; dim $I_g(M) = 0 \Leftrightarrow I_g(M)$ is a discrete group $\Leftrightarrow I_g^0(M) = \{\mathrm{id}_M\}$. When M is compact, we have additionally:

Proposition 2.2 *Let M be a compact connected n-manifold. Then the following are equivalent:*

1. *deg $M = 0$;*

2. *for each $g \in \mathcal{M}$, $I_g(M)$ is a finite group;*

3. *for each $g \in \mathcal{M}$, $\mathcal{I}_g(M) = 0$, i.e., there are no non-trivial Killing vector fields on M;*

4. *M does not admit an effective $\boldsymbol{SO}(2)$-action.*

Proof: $1 \Leftrightarrow 2$: If M is compact, then for any $g \in \mathcal{M}$, its isometry group $I_g(M)$ is compact. Thus if deg $M = 0$, then for all $g \in \mathcal{M}$, $I_g(M)$ is both compact and discrete and hence finite. Conversely, if $I_g(M)$ is finite for all $g \in \mathcal{M}$, then deg $M = 0$.

$2 \Leftrightarrow 3$: For M compact, $\mathcal{I}_g(M)$ is the Lie algebra of $I_g(M)$. Thus $I_g(M)$ is a finite group if and only if $\mathcal{I}_g(M) = 0$.

$1 \Rightarrow 4$: If M admits an effective $\boldsymbol{SO}(2)$-action, then averaging an arbitrary Riemannian metric $g \in \mathcal{M}$ with respect to this action yields an averaged metric $\bar{g}$ whose isometry group $I_{\bar{g}}(M) \supseteq \boldsymbol{SO}(2)$, and hence deg $M \geq$ dim $I_{\bar{g}}(M) \geq 1$, contradicting (1).

$4 \Rightarrow 1$: If deg $M \geq 1$, then there is a Riemannian metric $g \in \mathcal{M}$ such that dim $I_g(M) \geq 1$. Thus dim $I_g^0(M) =$ dim $I_g(M) \geq 1$, and since

M is compact, $I_g^0(M)$ is compact and connected. A maximal torus in $I_g^0(M)$ contains an effective $SO(2)$-action on M, contradicting (4). ∎

Recall that a *space form* is a complete connected Riemannian manifold of constant sectional curvature. The space form is *Euclidean* if the sectional curvature is zero, *spherical* if the sectional curvature is positive, and *hyperbolic* if the sectional curvature is negative.

The following definition is useful:

Definition 2.3 *A connected n-manifold M is of* **flat type** *(respectively,* **spherical type**, **hyperbolic type***) if M is diffeomorphic to a Euclidean space form (respectively, a spherical form, a hyperbolic space form).*

Note that in the above definition, we only require that M be diffeomorphic to a space form. We do not require that M have a Riemannian metric on it, nor if it does, need it be *isometric* to one of the space forms. Thus, for example, a 3-manifold is of spherical type if it is *diffeomorphic* to a spherical space form, i.e., a manifold of the form S^3/Γ, where Γ is a finite subgroup of $SO(4)$ acting freely and orthogonally (or isometrically) on S^3, here taken to be the standard 3-sphere of unit radius in R^4 (see Wolf [1972] for more information about space forms).

In three dimensions, the isometric classification of Euclidean space forms is known. We shall need the affine classification of the compact orientable case:

Theorem 2.4 *Let (M, g_F) be a compact connected orientable flat Riemannian 3-manifold. Then (M, g_F) is affinely diffeomorphic to one of six torsion-free affinely flat model spaces, which we denote by*

$\{\mathcal{F}_1, \ldots, \mathcal{F}_6\}$, where $\mathcal{F}_i = (F_i, \nabla_{F_i})$, $1 \leq i \leq 6$, and where F_i is the underlying manifold (of flat type) of $\mathcal{F}_i$, and ∇_{F_i} is the flat affine connection on F_i.

Here $\mathcal{F}_1$ is affinely diffeomorphic to any flat 3-torus $T^3 = R^3/(Z \times Z \times Z)$, and the other $\mathcal{F}_i$, $2 \leq i \leq 6$, are affinely diffeomorphic to Euclidean space forms R^3/Γ_i, where the Γ_i are described in Wolf [1972] (see also Ellis [1971] or Orlik [1972] for a description of the F_i). Moreover, an examination of the structure of these flat space forms shows that $\deg F_1 = \deg T^3 = 3$, $\deg F_i = 1$ for $2 \leq i \leq 5$, and that $\deg F_6 = 0$ (see also Remark 7 following Theorem 4.1).

Now we recall the following regarding Eilenberg-MacLane spaces (see for example Spanier [1966]).

Definition 2.5 *Let i be an integer ≥ 1 and let π be a group, abelian if $i \geq 2$. A topological path-connected pointed space (X, x_0) is an* **Eilenberg-MacLane space of type** (π, i)*, or a $K(\pi, i)$-space, if the ith homotopy group $\pi_i(X, x_0) = \pi$, and $\pi_j(X, x_0) = 0$ for $j \neq i$.*

Thus $K(\pi, i)$ topological spaces are spaces all of whose homotopy is concentrated in a single group. The usefulness of such spaces is that their homotopy is relatively simple, and thus they can be used as a set of building blocks for understanding more complicated spaces.

For our purposes, the $K(\pi, 1)$-spaces will be the most important. In fact, we make the following definition:

Definition 2.6 *A connected manifold M is a $K(\pi, 1)$-**manifold**, or is* **aspherical***, if M is a $K(\pi, 1)$-space.*

The term *aspherical* arises because the homotopy groups $\pi_i(M)$, $i \geq 1$, can be thought of as homotopy classes of continuous maps from spheres S^i into M. Thus if M is aspherical, all such continuous maps for $i > 1$ are null homotopic, i.e., are homotopic to the constant map. In particular, any embedding of an i-sphere, $i > 1$, into M is null homotopic, and hence any embedded i-sphere, $i > 1$, can be contracted to a point. In this sense, M is aspherical.

A nice class of examples of $K(\pi, 1)$-manifolds is given by the following:

Proposition 2.7 *Let M be a manifold of either flat or hyperbolic type. Then M is a $K(\pi, 1)$-manifold.*

Proof: If M is of flat type, then from the classical theorem of Killing and Hopf (see Wolf [1972]), M is diffeomorphic to R^n/Γ where Γ is a subgroup of the Euclidean group acting freely and properly discontinuously on R^n. Since R^n is contractible and since the orbit projection map $R^n \to R^n/\Gamma$ is a covering map, $\pi_1(R^n/\Gamma) = \Gamma$. Since the homotopy sequence of a fibration is exact, $\pi_i(R^n/\Gamma) = \pi_i(R^n) = 0$ for $i \geq 2$. Thus $M \approx R^n/\Gamma$ is a $K(\Gamma, 1)$-manifold.

If M is of hyperbolic type, then M is diffeomorphic to H^n/Γ, where H^n is unit hyperbolic space, and where Γ is a subgroup of the isometry group $I(H^n)$ acting freely and properly discontinuously on H^n. Since H^n is diffeomorphic to R^n, the same argument applies. ∎

More generally, a manifold M is a $K(\pi, 1)$-manifold if and only if the universal cover of M is a contractible manifold.

The connected sum $M_1 \# M_2$ of two connected orientable n–ma ni folds M_1 and M_2 is defined by removing the interior of a solid n-ball from each and then matching the resulting boundaries using an orientation-reversing diffeomorphism (e.g., see Bröcker and Jänich [1982]). The resulting manifold does not depend on which balls are removed. This operation is well-defined up to diffeomorphism, is associative and commutative (up to diffeomorphism), and the n-sphere S^n serves as an identity element.

If two groups G_1 and G_2 are isomorphic, we denote this by $G_1 \approx G_2$. If two manifolds M_1 and M_2 are diffeomorphic, we also denote this by $M_1 \approx M_2$. A connected n-manifold M is *trivial* if $M \approx S^n$ and is *non-trivial* if $M \not\approx S^n$.

Definition 2.8 *Let M be a non-trivial connected orientable n–ma ni fold M. M is* **decomposable** *(or* **non-prime**, *or* **composite***) if there exists a* **decomposition** $M \approx M_1 \# M_2$, *where M_1 and M_2 are both non-trivial connected orientable n-manifolds. M is* **prime** *if there does not exist any decomposition of M.*

A related notion is that of irreducible:

Definition 2.9 *A compact connected orientable 3-manifold is* **irreducible** *if every 2-sphere in M bounds a 3-cell.*

Let M be a compact connected orientable 3-manifold. With the exception of S^3 (which is irreducible but not prime) and $S^1 \times S^2$ (which is prime but not irreducible), M is prime if and only if M is irreducible (Milnor [1961]).

Now we recall the prime decomposition theorem for compact connected orientable 3-manifolds.

Theorem 2.10 (Kneser (1929), Milnor (1961)) *Let M be a non-trivial compact connected orientable 3-manifold. Then M is diffeomorphic to a finite connected sum of compact prime manifolds*

$$M \approx P_1 \# \cdots \# P_k \,,$$

where the summands are uniquely determined up to order and diffeomorphism.

The existence part of the above Theorem was proved by Kneser [1929] and the uniqueness part was proved by Milnor [1961].

Now recall that a *homotopy 3-sphere* is a compact connected topological 3-manifold which is homotopic to a 3-sphere. A homotopy 3-sphere may or may not be homeomorphic to a 3-sphere. A *fake 3-sphere* is a homotopy 3-sphere which is not homeomorphic to a 3-sphere. Similarly, a *homotopy 3-cell* is a 3-cell that is homotopic to the closed unit ball B^3 in R^3, and a *fake 3-cell* is a homotopy 3-cell that is not homeomorphic to B^3. Poincaré's conjecture asserts that any compact connected simply connected topological 3-manifold is homeomorphic to a 3-sphere. Poincaré's conjecture is equivalent to the non-existence of fake 3-spheres which in turn is equivalent to the non-existence of fake 3-cells (see e.g. Hempel [1976]).

We also remark that every topological 3-manifold has a differentiable structure unique up to diffeomorphism (Munkres [1960] and Whitehead [1961]), so that for considerations regarding 3-manifold topology, we may work in the differentiable category without loss of generality.

If M is prime, then M is either finitely covered by an irreducible homotopy 3-sphere Σ^3, in which case its fundamental group $\pi_1(M) \approx \Gamma$ is a finite subgroup of $\mathcal{D}^+(M)$ acting freely on M, or M is diffeomorphic to a handle $S^1 \times S^2$ (also known as a wormhole), or M is an irreducible $K(\pi, 1)$-manifold, in which case its fundamental group $\pi_1(M) \approx \pi$ is infinite and torsion-free (i.e. no non-trivial elements of π have finite order); see Milnor [1961]. Thus Theorem 2.10 can be refined as follows:

Theorem 2.11 (Prime Decomposition Theorem) *Let M be a non-trivial compact connected orientable 3-manifold. Then M is diffeomorphic to a finite connected sum of compact connected orientable* *prime 3-manifolds of the form*

$$M \approx \underbrace{\Sigma_1^3/\Gamma_1 \# \ldots \# \Sigma_k^3/\Gamma_k}_{\text{spherical-type factors}} \# \underbrace{(S^1 \times S^2)_1 \# \ldots \# (S^1 \times S^2)_l}_{\text{handles (or wormholes)}}$$

$$\# \underbrace{K(\pi_1, 1) \# \ldots \# K(\pi_m, 1)}_{K(\pi,1)\text{-factors}}$$

where

1. *k, l, and m are non-negative integers, $k + l + m \geq 1$ (if $k + l + m = 1$, then M is prime), and if any of the integers k, l, or m are zero, then terms of that type do not appear;*

2. *if $k \geq 1$, then for each i, $1 \leq i \leq k$,*

   (a) *Σ_i^3 is an irreducible homotopy 3-sphere (which may be diffeomorphic to S^3, in which case $\Gamma_i \neq \{\mathrm{id}_{\Sigma_i^3}\}$; see (c));*

   (b) *$\Gamma_i \subset \mathcal{D}^+(\Sigma_i^3)$ is a finite subgroup of orientation-preserving diffeomorphisms of Σ_i^3 acting freely on Σ_i^3, so that $\pi_1(\Sigma_i^3/\Gamma_i) \approx \Gamma_i$;*

   (c) *no factor Σ_i^3/Γ_i is trivial ($\approx S^3$); thus if $\Sigma_i^3 \approx S^3$, then Γ_i is non-trivial ($\neq \{\mathrm{id}_{\Sigma_i^3}\}$), whereas if the Poincaré conjecture is false and $\Sigma_i^3 \not\approx S^3$, then the group Γ_i may be trivial ($= \{\mathrm{id}_{\Sigma_i^3}\}$);*

3. *if $m \geq 1$, then for each j, $1 \leq j \leq m$, the factor $K(\pi_j, 1)$ is a compact orientable irreducible $K(\pi_j, 1)$-manifold whose fundamental group $\pi_1(K(\pi_j, 1)) = \pi_j$ is infinite and torsion-free;*

and where the summands in Eqn (2-1) are uniquely determined up to order and diffeomorphism.

By the Seifert-Van Kampen Theorem (see e.g. Massey [1967]), the fundamental group of the connected sum of two n-manifolds,

$n \geq 3$, is given by $\pi_1(M_1 \# M_2) = \pi_1(M_1) * \pi_1(M_2)$, where $*$ denotes the free product of groups. In particular, the connected sum of two irreducible fake 3-spheres (if such exist), $\Sigma^3_{f_1} \# \Sigma^3_{f_2} = \Sigma^3_{f_3}$ (with possibly $\Sigma^3_{f_1} \approx \Sigma^3_{f_2}$), has trivial fundamental group and thus is a non-irreducible fake 3-sphere. Thus if the Poincaré conjecture is false, there exist non-irreducible fake 3-spheres.

Similarly, since up to homotopy, a fake 3-sphere acts like a true 3-sphere, any connected sum $\Sigma^3_f \# M$ of a fake 3-sphere Σ^3_f and a 3-manifold M is homotopy equivalent to M; thus, up to homotopy, a fake 3-sphere like a true 3-sphere is invisible in a connected sum. In particular, the connected sum $\Sigma^3_f \# K'(\pi, 1)$ of a fake 3-sphere Σ^3_f and an irreducible $K'(\pi, 1)$-manifold is homotopy equivalent to $K'(\pi, 1)$ and thus is a non-irreducible $K(\pi, 1)$-manifold with the same π. Equivalently, if a $K(\pi, 1)$-manifold M contains a fake 3-cell, then M can be expressed as a connected sum of a fake 3-sphere and a $K'(\pi, 1)$-manifold. Thus if the Poincaré conjecture is false, there exist non-irreducible $K(\pi, 1)$-manifolds. Conversely, if the Poincaré conjecture is true, then every $K(\pi, 1)$-manifold is irreducible (Milnor [1961]). Thus every $K(\pi, 1)$-manifold is irreducible if and only if the Poincaré conjecture is true.

Since the Prime Decomposition Theorem does not assume that the Poincaré conjecture is true, in (2a) above we must restrict to irreducible homotopy 3-spheres, and in (3) we must restrict to irreducible $K(\pi, 1)$-manifolds.

Since M is orientable, each factor in the prime decomposition of M must be orientable. Now each spherical-type factor Σ^3_i/Γ_i is orientable if and only if Γ_i is a subgroup of $\mathcal{D}^+$, so that in (2b), each Γ_i, $1 \leq i \leq k$, is required to be a finite subgroup of $\mathcal{D}^+$.

We remark again that M is only diffeomorphic to such a connected sum, and that M is not endowed with any Riemannian metric.

Now suppose that the Poincaré conjecture is true. Then every homotopy 3-sphere is a true 3-sphere, so that in this case each spherical-type factor Σ_i^3/Γ_i in Eqn (2–1) reduces to a spherical-factor of the form S^3/Γ_i, where $\Gamma_i \subset \mathcal{D}^+(S^3)$ is a finite non-trivial group of orientation-preserving diffeomorphisms of S^3 acting freely. If this action is equivalent (i.e., conjugate in $\mathcal{D}^+(S^3)$) to an orthogonal action of S^3, then, up to diffeomorphism, Γ_i can be taken to be a finite non-trivial subgroup of $SO(4)$ acting freely and orthogonally on S^3. In this case, the resulting orbit space S^3/Γ_i is a non-trivial orientable spherical space form.

However, there is a question of *whether or not every free action of a finite group acting on S^3 is equivalent to a standard orthogonal action* (see the reviews by Davis and Milgram [1985], Edmonds [1985], and the references therein). If such is the case, then every 3-manifold whose universal cover is S^3 is diffeomorphic to a spherical space form. For our purposes, we shall not need to assume that such is the case (however, see the remarks following Theorem 4.3 and Definition 8.1).

If the Poincarè conjecture is true, then Eqn 2–1 of Theorem 2.11 can be replaced by the following:

$$M \;\approx\; \underbrace{S^3/\Gamma_1 \# \ldots \# S^3/\Gamma_k}_{\text{spherical-factors}} \# \underbrace{(S^1 \times S^2)_1 \# \ldots \# (S^1 \times S^2)_l}_{\text{handles (or wormholes)}}$$
$$\# \underbrace{K(\pi_1, 1) \# \ldots \# K(\pi_m, 1)}_{K(\pi,1)\text{-factors}}$$

still subject to (1) and (3), but where (2a,b,c) is simplified to:

(2′) *If $k \geq 1$, then for each i, $1 \leq i \leq k$, $\Gamma_i \subset \mathcal{D}^+(S^3)$ is a finite non-trivial subgroup of orientation-preserving diffeomorphisms of S^3 acting freely on S^3, so that $\pi_1(S^3/\Gamma_i) \approx \Gamma_i$.*

Finally, we remark that if the Poincaré conjecture is true, a $K(\pi, 1)$–manifold is automatically irreducible, so that in this case the stated irreducible condition on the $K(\pi, 1)$-factors is automatic.

Convention 2.12 We shall be frequently considering connected sums of the type of Eqns (2–1) and (2–2). Therefore, whenever we write such a connected sum, we shall adopt the convention that any connected sum of the type of Eqn (2–1) will always be subject to the conditions (1), (2a,b,c), and (3) of Theorem 2.11, and that any connected sum of the type of Eqn (2–2) will always be subject to conditions (1), (2′), and (3). Any *additional* restrictions for either connected sum will be stated explicitly.

3   Homotopically trivial symmetries of most 3-manifolds are toral

In this section we describe the recent work of Mess [1995] (see also Freedman and Yau [1983]). We need the following terminology:

Definition 3.1 *Let M be a connected n-manifold. A **symmetry** of M is a diffeomorphism $f \in \mathcal{D}$ of finite order. A non-trivial symmetry $f (\neq id_M)$ of M is **toral** if the action of the finite cyclic group $\langle f \rangle$ generated by f embeds in an effective action of $SO(2)$ on M.*

The case of interest to us is $n = 3$. In this case, we remark that for Freedman and Yau [1983], in a similar setting, toral refers to an effective action of either $S^1 = SO(2)$, $S^1 \times S^1$, or $S^1 \times S^1 \times S^1$.

Theorem 3.2 (Mess [1995]) *Let M be a compact connected irreducible aspherical 3-manifold. Let $f \in \mathcal{D}$ be a non-trivial symmetry of M that is homotopic to the identity id_M. Then f is toral.*

If M is not irreducible, then there is an $\langle f \rangle$-equivariant collection of mutually disjoint fake 3-cells such that after replacing each fake 3-cell by a standard 3-cell, f becomes toral. Moreover, the stabilizer in $\langle f \rangle$ of each fake 3-cell in the collection is trivial.

Our case of interest will be when M is orientable and when the the results of Theorem 3.2 hold without replacing fake 3-cells by standard 3-cells. In light of the Prime Decomposition Theorem 2.11, a weaker form of Mess' Theorem can be restated as follows:

Theorem 3.3 *Let M be a compact connected orientable 3-manifold. Assume that either*

(1) M is an irreducible $K(\pi, 1)$-manifold; or
(2) M is a composite manifold diffeomorphic to a connected sum of the form of Eqn 2-2; namely,

$$M \approx \underbrace{S^3/\Gamma_1 \# \ldots \# S^3/\Gamma_k}_{spherical\text{-}factors} \# \underbrace{(S^1 \times S^2)_1 \# \ldots \# (S^1 \times S^2)_l}_{handles \ (or \ wormholes)}$$
$$\# \underbrace{K(\pi_1, 1) \# \ldots \# K(\pi_m, 1)}_{K(\pi,1)\text{-}factors}$$

where conditions (1), (2'), and (3) of Theorem 2.11 hold, and where $k + l + m \geq 2$ (since M is a composite manifold).

Then every non-trivial symmetry of M homotopic to the identity is toral.

Proof: (1) An irreducible $K(\pi, 1)$-manifold is the same as an irreducible aspherical manifold.

(2) Any composite 3-manifold is not irreducible. Since the spherical-factors S^3/Γ_i and the $K(\pi, 1)$-factors in the composite manifold given by Eqn 3-1 are irreducible, they cannot contain a fake 3-cell,

and thus M cannot contain a fake 3-cell (see also the remarks following Theorem 2.11). Thus Theorem 3.2 applied to M shows that any non-trivial symmetry of M homotopic to the identity is toral. ∎

4   Three-manifolds with $\deg M = 0$

From Proposition 2.2, we have seen that for a compact connected n-manifold M, $\deg M = 0$ if and only if M does not admit an effective $SO(2)$-action. For our purposes, the case of compact 3-manifolds are of most importance, inasmuch as these manifolds play a fundamental role in the dynamical formulation of general relativity. Fortunately, the results of Raymond [1968] and Orlik and Raymond [1968] give a complete topological and equivariant classification of those compact 3-manifolds that can support an effective $SO(2)$-action. Here we are only interested in the topological classification and the orientable case, but the equivariant classification and the non-orientable case are treated there as well.

The manifolds of most interest to us are those of $\deg M = 0$. These manifolds will then be the manifolds that are not in their list. From the prime decomposition theorem for 3-manifolds, these excluded manifolds, in turn, can then be listed. Again we remark that those 3-manifolds with $\deg M = 0$, although not the most familiar ones, are the generic ones, and in this sense are worthy of the most attention.

First we describe the Orlik-Raymond list, and then describe the complement to their list. Their list for the orientable case (without the actions) is the following:

Theorem 4.1 (Orlik-Raymond (1968)) *Let M be a compact con-*

nected orientable 3-manifold such that M admits an effective $SO(2)$-action. Then M is diffeomorphic to one of the following manifolds:

1. *S^3, $S^1 \times S^2$, or a lens space $L(p,q)$ (see Remark 2 below).*

2. *A connected sum of the above (see Remark 3).*

3. *A quotient of $SO(3)$ or $Spin(1)$ by a finite, non-abelian, discrete subgroup (see Remark 4).*

4. *A $K(\pi,1)$-manifold whose fundamental group π has infinite cyclic center (provided it is not the 3-dimensional torus) (see Remarks 5, 6, and 7).*

Remarks:

1. See Fischer ([1970]) for an application of these results to the construction of superspace, where also the degree of symmetry of each of the above manifolds is given.

2. In the construction of the lens spaces $L(p,q) = S^3/\Gamma(p,q)$, $0 < q < p$ and $(p,q) = 1$ (p, q relatively prime), S^3 is viewed as the unit quaternions, i.e., the unit sphere in $C \times C$, with multiplicative group structure induced from $C \times C$ (making S^3 isomorphic to $SU(2) = Spin(1)$). The group $\Gamma(p,q) \subset SO(2) \times SO(2) \subset SO(4)$ denotes the cyclic group of order p generated by $(e^{2\pi i/p}, e^{2\pi i q/p})$, acting on the unit sphere in $C \times C$. Moreover, every free orthogonal action of $Z_p = Z/pZ$ on S^3 is equivalent to some action of $\Gamma(p,q)$ on S^3. Note that with the above restrictions on p, q, S^3 is not considered a lens space, whereas projective 3-space $P^3 = L(2,1)$ is the "first" lens space.

3. Since S^3 acts as the identity with respect to connected sums, their item (2) can be expressed as a connected sum of lens

spaces and handles. Note that such a connected sum is the only composite manifold that can admit an $SO(2)$-action.

4. Their item (3) above must be interpreted as "M is a quotient of S^3 by a finite non-abelian subgroup Γ of $SO(4)$ acting freely and orthogonally on S^3", since such manifolds are spherical space forms and admit $SO(2)$-actions (see Witt [1986] where the isometry groups of the spherical space forms are calculated). Thus the manifolds that occur in their item (3) are the non-trivial orientable spherical space forms that are not lens spaces. The lens spaces are included in item (1) so as to be able to be part of the connected sums in item (2). Also note that the only finite abelian subgroups of $SO(4)$ are cyclic groups, and thus the lens spaces are the only spherical space forms that are quotients of S^3 by a finite abelian subgroup of $SO(4)$ acting freely and orthogonally.

5. Their item (4) above must be interpreted as "M is diffeomorphic to either the 3-torus T^3, or to a compact orientable irreducible $K(\pi, 1)$-manifold whose fundamental group π has infinite cyclic center", so that the 3-torus is included in this item. The 3-torus is considered separately because it is a $K(\pi, 1)$-manifold whose fundamental group $\pi_1(T^3) = Z \oplus Z \oplus Z$ does not have infinite cyclic center. The $K(\pi, 1)$-manifold must be irreducible so that a connected sum of a fake 3-sphere and an irreducible $K(\pi, 1)$-manifold that admits a circle action is excluded (see Orlik [1972, pp. 19 and 21] and also the remarks after Theorem 2.11).

6. In their item (4), the structure of the $K(\pi, 1)$-manifolds M that can occur are described more fully by considering the orbit space $M/SO(2)$ and the orbit projection map $\pi : M \to M/SO(2)$. Their results are that $\pi : M \to M/SO(2)$ is a

Seifert fibered space, and the action and the topology of M are determined uniquely by the numerical Seifert invariants of M. Moreover, if $M \not\approx T^3$, then the center of $\pi = \pi_1(M)$ is infinite cyclic and is generated by an orbit of the action; i.e., there exists an orbit $\gamma \subset M$ of the action whose homotopy class $[\gamma] \in \pi_1(M)$ generates the cyclic group $\langle [\gamma] \rangle \approx \text{Center}(\pi_1(M)) \approx \mathbf{Z}$. However, it is not known if the converse is true. Thus there *may* exist $K(\pi, 1)$-manifolds with $\text{Center}(\pi) \approx \mathbf{Z}$ but which do not admit an $\mathbf{SO}(2)$ action. By the results of Waldhausen ([1967], [1968]), such a manifold must be a non-Haken $K(\pi, 1)$-manifold (see Section 5).

7. Now further partition the irreducible $K(\pi, 1)$-manifolds into those of flat types and those of non-flat types. Examining the topology of the flat types $\{F_1, F_2, F_3, F_4, F_5, F_6\}$ ($F_1 = \mathbf{T}^3$) shows that only F_6 does not admit an $\mathbf{SO}(2)$-action (see Ellis [1971], Wolf [1972], Orlik [1972], and also the remarks following Theorem 2.4). Thus their item (4) can be expressed as follows: *Either M is of flat type diffeomorphic to F_1, F_2, F_3, F_4, or F_5 (F_6 is excluded), or M is a compact orientable $K(\pi, 1)$-manifold of non-flat type with* $\deg M = 1$. That the degree of this latter class of manifolds cannot exceed 1 follows from an examination of those manifolds that have degree ≥ 2 (see Fischer [1970] and the references cited therein). Thus the compact orientable irreducible $K(\pi, 1)$-manifolds with $\deg > 0$ consist of $F_1 = \mathbf{T}^3$ with $\deg \mathbf{T}^3 = 3$, the flat types F_2, F_3, F_4, and F_5, each with degree $= 1$, and the non-flat types with degree $= 1$. This further refinement of their item (4) is useful when analyzing the initial value problem in general relativity, as it is natural to divide the $K(\pi, 1)$-manifolds into those of flat and non-flat type (see Fischer-Moncrief ([1994a,b], [1995a,b])).

With these remarks, the description of the deg $M > 0$ manifolds can be re-expressed as follows:

Theorem 4.2 *Let M be a compact connected orientable 3-manifold with $\deg M > 0$. Then either (1) or (2) hold:*

1. *$M \approx S^3$; or M is diffeomorphic to either*

   (a) *a non-trivial orientable spherical space form S^3/Γ, where Γ is a finite non-trivial subgroup of $SO(4)$ acting freely and orthogonally on S^3; in this case, either*

      i. *Γ is cyclic and non-trivial, so that $\Gamma \approx \Gamma(p,q)$ for some p, q, $0 < q < p$, $(p,q) = 1$, and M is a lens space, $M \approx L(p,q) = S^3/\Gamma(p,q)$; or*

      ii. *Γ is non-abelian so $M \approx S^3/\Gamma$ is a non-trivial orientable spherical space form that is not a lens space;*

   (b) *a handle $S^1 \times S^2$;*

   (c) *one of the following $K(\pi, 1)$-manifolds:*

      i. *$F_1(= T^3)$, F_2, F_3, F_4, or F_5; or*

      ii. *a compact orientable irreducible $K(\pi, 1)$-manifold M of non-flat type such that $\deg M = 1$; in this case, $M \to M/SO(2)$ is a Seifert fibered space determined by its numerical Seifert invariants, and $\mathrm{Center}(\pi) \approx \mathbf{Z}$.*

2. *M is a composite manifold and is diffeomorphic to a finite connected sum of lens spaces and handles*

$$M \approx L(p_1, q_1) \# \ldots \# L(p_k, q_k) \# (S^1 \times S^2)_1 \# \ldots \# (S^1 \times S^2)_l \tag{4-1}$$

where $k, l \geq 0$, $k + l \geq 2$ (since M is composite), and where if $k \geq 1$, then for each i, $1 \leq i \leq k$, $0 < q_i < p_i$ and $(p_i, q_i) = 1$.

The Orlik-Raymond Theorem 4.1 results list the manifolds M with $\deg M > 0$. Now we consider the complement to their list, namely, those M with $\deg M = 0$. Using the alternate description above (Theorem 4.2) allows for the following description:

Theorem 4.3 *Let M be a compact connected orientable 3-manifold. Then $\deg M = 0$ if and only if either (1) or (2) holds:*

1. *M is a prime manifold and is diffeomorphic to one of the following:*

   (a) *a manifold Σ^3/Γ, where Σ^3 is a homotopy 3-sphere, $\Gamma \subset \mathcal{D}^+(\Sigma^3)$ is a finite group of orientation-preserving diffeomorphisms of Σ^3 acting freely on Σ^3, and such that Σ^3/Γ is not diffeomorphic to a spherical space form S^3/Γ', where Γ' is a finite subgroup of $SO(4)$ acting freely and orthogonally on S^3;*

   (b) *the compact orientable flat space form F_6;*

   (c) *a compact orientable irreducible $K(\pi, 1)$-manifold M of non-flat type such that $\deg M = 0$; a sufficient condition for this case is that $\mathrm{Center}(\pi) \not\approx Z$ (see Remark 6 following Theorem 4.1).*

2. *M is a composite manifold of the form of Eqn 2–1*

$$M \approx \Sigma_1^3/\Gamma_1 \# \ldots \# \Sigma_k^3/\Gamma_k \# (S^1 \times S^2)_1 \# \ldots \# (S^1 \times S^2)_l$$
$$\# \; K(\pi_1, 1) \# \ldots \# K(\pi_m, 1)$$

   where conditions (1), (2a,b,c), and (3) of Theorem 2.11 hold, and where $k + l + m \geq 2$ (since M is composite), and such that either

   (a) *$m \geq 1$ (so that at least one of the compact orientable irreducible $K(\pi_j, 1)$-factors occurs); or*

(b) *if* $m = 0$, *then* $k \geq 1$ *(so that at least one of the spherical-type factors* Σ_i^3/Γ_i *occurs), and such that at least one of these factors is not diffeomorphic to a lens space.*

Proof: If M is diffeomorphic to any of the manifolds in (1a,b,c) or (2a,b), then M does not appear on the Orlik-Raymond list, and so $\deg M = 0$.

Conversely, first assume that M is prime, so that M is diffeomorphic to either

1. Σ^3/Γ where Σ^3 is an irreducible homotopy 3-sphere (which may be S^3) and $\Gamma \subset \mathcal{D}^+(\Sigma^3)$ is a finite group of orientation-preserving diffeomorphisms of Σ^3 acting freely on Σ^3, where Γ is non-trivial if $\Sigma^3 \approx S^3$;

2. a handle $S^1 \times S^2$; or

3. a compact orientable irreducible $K(\pi, 1)$-manifold.

Now assume additionally that $\deg M = 0$. Since an orientable spherical space form S^3/Γ (where Γ is a finite subgroup of $SO(4)$ acting freely and orthogonally on S^3) or a handle $S^1 \times S^2$ have $\deg M > 0$, item (2) is ruled out, and if M is of the form (1) Σ^3/Γ, then Σ^3/Γ cannot be diffeomorphic to an orientable spherical space form (since such have degree > 0).

If M is a compact orientable irreducible $K(\pi, 1)$-manifold of flat type, then M is diffeomorphic to F_6. Otherwise M is of non-flat type with $\deg M = 0$; a sufficient condition for this is that Center$(\pi) \not\approx \mathbf{Z}$ (see Remark 6 following Theorem 4.1).

Now assume M is composite. From the Orlik-Raymond list, the only composite manifolds with $\deg M > 0$ are connected sums of handles and lens spaces. Thus if either M has a compact orientable irreducible $K(\pi, 1)$-factor in its prime decomposition (case (2a)), or if M does not have a compact orientable irreducible $K(\pi, 1)$-factor in its prime decomposition, but does have a spherical-factor Σ^3/Γ that is not a lens space (case (2b)), then deg $M = 0$. ∎

Theorems 4.2 and 4.3 are complementary in the sense that the manifolds described in each Theorem are mutually exclusive and exhaust the possible classes of 3-manifolds. Thus Theorems 4.2 and 4.3 classify compact connected orientable 3-manifolds according to whether their degree of symmetry is positive or zero.

If we assume that the Poincaré conjecture is true, then the description of (1a) and Eqn 4–2 of Theorem 4.3 can be simplified to:

(1a′) *A manifold S^3/Γ, where $\Gamma \subset \mathcal{D}^+(S^3)$ is a finite group acting freely such that S^3/Γ is not diffeomorphic to an orientable spherical space form* (if such exist; see below). Moreover, in this case, Eqn 4–2 can be replaced by

$$M \approx S^3/\Gamma_1 \# \ldots \# S^3/\Gamma_k \# (S^1 \times S^2)_1 \# \ldots \# (S^1 \times S^2)_l$$
$$\# K(\pi_1, 1) \# \ldots \# K(\pi_m, 1)$$

together with conditions (2a,b) of Theorem 4.3.

Note that if we also assume that the conjecture that every finite free action on S^3 is equivalent to a standard orthogonal action is true (see the remarks after Theorem 2.11), then there are no manifolds of the type (1a′). Thus under this circumstance, this case (1a′) does not occur.

Note also that although the description of the deg $M = 0$ manifolds is simplified if we assume that the Poincaré conjecture is true, this assumption does not change the description of the deg $M > 0$ manifolds (Theorem 4.2) inasmuch as neither a fake 3-sphere nor a manifold of the form S^3/Γ that is not a spherical space form can admit an $SO(2)$-action. Thus if such manifolds exist, their degree is known, and it is zero.

5 Contractibility of $\mathcal{D}_0$

In this section we discuss the work of Hatcher [1976] and Waldhausen ([1967], [1968]) that we shall need. To discuss this work, we need some preliminary definitions. Let M be a compact connected orientable 3-manifold, let S be a compact connected orientable 2-manifold, and let

$$i : S \to M$$

be an embedding of S into M so that $i(S)$ is an embedded surface in M. Then i induces a homomorphism on the homotopy groups

$$i_* : \pi_j(S) \to \pi_j(M) \, , \, j \geq 1 \, .$$

The embedded surface $i(S)$ is *incompressible* if the induced homomorphism i_* is injective on the fundamental groups, $i_* : \pi_1(S) \to \pi_1(M)$. In other words, non-contractible loops in the embedded surface $i(S) \subset M$ are still non-contractible even when allowed to move off of the embedded surface $i(S)$ in the ambient space M.

A 3-manifold is *sufficiently large* if it contains an incompressible surface S of genus ≥ 1. In particular, the fundamental group of such a manifold must contain as a subgroup the fundamental group of such a surface, and thus must be infinite $\not\approx \mathbf{Z}$. A *Haken manifold* M is an irreducible compact connected orientable sufficiently large 3-manifold. Since a Haken manifold is irreducible and since its fundamental group is infinite $\not\approx \boldsymbol{Z}$, a Haken manifold must be an irreducible $K(\boldsymbol{\pi}, 1)$-manifold.

Haken manifolds share many of their properties with closed orientable surfaces ($\not\approx \boldsymbol{S}^2$). For one thing, both are examples of $K(\boldsymbol{\pi}, 1)$-manifolds. For another, there is a hierarchy structure of Haken manifolds which is analogous to a sequence of cuts turning a surface into a disk.

Hatcher's Theorem gives information about the homotopy groups of $\mathcal{D}$ when M is a Haken manifold. For a group G, let $\mathrm{Aut}(G)$ denote the group of automorphisms of G, let $\mathrm{In}(G)$ denote the normal subgroup of inner automorphisms, and let $\mathrm{Out}(G) = \mathrm{Aut}(G)/\mathrm{In}(G)$ denote the quotient group of outer automorphisms. Let $\mathrm{Center}(G)$ denote the center of G.

Theorem 5.1 (Hatcher [1976]) *Let M be a Haken 3-manifold with fundamental group $\pi_1(M) \approx \boldsymbol{\pi}$. Then*

$$
\begin{aligned}
\pi_0(\mathcal{D}) &\approx \mathcal{D}/\mathcal{D}_0 &\approx \mathrm{Out}(\boldsymbol{\pi}) \\
\pi_1(\mathcal{D}) &\approx \pi_1(\mathcal{D}_0) &\approx \mathrm{Center}(\boldsymbol{\pi}) \\
\pi_i(\mathcal{D}) &\approx \pi_i(\mathcal{D}_0) &= 0 \; for \; i \geq 2
\end{aligned}
$$

The proof is given in Hatcher [1976]; the above formulation is given in Giulini [1995]. That $\pi_i(\mathcal{D}) \approx \pi_i(\mathcal{D}_0)$ for $i \geq 1$ follows from the fact that the homotopy groups of a disconnected space are given by the homotopy groups of the component in which its base point is located.

From the results of Raymond and Orlik (see Section 4), if M is a Seifert fibered manifold that admits an $\boldsymbol{SO}(2)$-action, then either $M \approx \boldsymbol{T}^3$, or $\mathrm{Center}(\pi_1(M)) \approx \boldsymbol{Z}$ and is generated by an orbit of the action (see Remark 6 following Theorem 4.1). Waldhausen has the following converse:

Theorem 5.2 (Waldhausen [1967], [1968]) *Let M be a Haken 3-manifold with fundamental group $\pi_1(M) \approx \pi$. If* $\mathrm{Center}(\pi) \neq 0$, *then M is a Seifert fibered manifold, in which case either $M \approx T^3$, or* $\mathrm{Center}(\pi) \approx \mathbf{Z}$ *and is generated by an orbit of an effective $\mathbf{SO}(2)$-action on M.*

Now if M is a Haken manifold, then in particular, M is an irreducible $K(\pi, 1)$-manifold. It is reasonable to conjecture that the Haken condition is not critical, and that this result is true more generally for M a compact connected orientable irreducible $K(\pi, 1)$-manifold. If this were the case, then such a $K(\pi, 1)$-manifold $(\not\approx T^3)$ would have an effective $\mathbf{SO}(2)$-action if and only if $\mathrm{Center}(\pi) \approx \mathbf{Z}$.

Putting together Theorems 5.1 and 5.2, we get the following:

Theorem 5.3 (Contractibility of $\mathcal{D}_0$) *Let M be a Haken 3-manifold such that $\deg M = 0$. Then $\mathcal{D}_0$ is contractible.*

Proof: Let $\pi_1(M) = \pi$. From Theorem 5.1, $\pi_1(\mathcal{D}_0) \approx \mathrm{Center}(\pi)$ and $\pi_i(\mathcal{D}_0) = 0$ for $i > 1$. From Theorem 5.2, if $\mathrm{Center}(\pi) \neq 0$, then M is a Seifert fibered manifold, in which case either $M \approx T^3$, or $\mathrm{Center}(\pi) \approx \mathbf{Z}$ and is generated by an orbit of an effective $\mathbf{SO}(2)$-action on M. In either case, $\deg M > 0$, contradicting $\deg M = 0$. Thus $\pi_1(\mathcal{D}_0) \approx \mathrm{Center}(\pi) = 0$. Thus $\pi_i(\mathcal{D}_0) = 0$ for $i \geq 1$, so that $\mathcal{D}_0$ is weakly homotopic to a point. Now $\mathcal{D}_0$ is an ILH-manifold and hence a metrizable manifold (see e.g. Fischer [1970]). But metrizable manifolds are weakly homotopic to a point if and only if they are contractible (see Palais [1966]). Thus $\mathcal{D}_0$ is contractible. ∎

If Theorem 5.2 turns out to be true more generally for (non-Haken) irreducible $K(\pi, 1)$-manifolds, then $\mathcal{D}_0$ would also be contractible for these manifolds. Indeed, if M is a $K(\pi, 1)$-manifold, M is covered by a contractible manifold, and it is plausible that the contractibility of the universal cover of M would be reflected by a similar contractibility in the structure of $\mathcal{D}_0$.

6   Classical and quantum superspace

Let M be a compact connected n-manifold. The group $\mathcal{D}$ acts on $\mathcal{M}$ on the right by pull-back

$$\Phi : \mathcal{M} \times \mathcal{D} \longrightarrow \mathcal{M}; \quad (g, f) \longmapsto f^*g$$

For $g \in \mathcal{M}$, we let

$$\mathcal{O}_g = g \cdot \mathcal{D} = \{ f^*g \mid f \in \mathcal{D} \}$$

denote the orbit through g, so that $\mathcal{O}_g$ is the set of metrics in $\mathcal{M}$ that are isometric to g.

Classical superspace (or *superspace*) is defined as the orbit space

$$\mathcal{S} = \mathcal{M}/\mathcal{D} = \{ [g] \mid g \in \mathcal{M} \}$$

where $[g]$ denotes the orbit of g taken as a point in $\mathcal{M}/\mathcal{D}$ (as opposed to the orbit $\mathcal{O}_g$, a subset of $\mathcal{M}$). Let

$$\pi : \mathcal{M} \longrightarrow \mathcal{M}/\mathcal{D}; \quad g \longmapsto [g]$$

denote the projection onto the orbit space, so that $\mathcal{O}_g = \pi^{-1}([g])$. Classical superspace is taken in the quotient, or orbit space topology, where a subset $\mathcal{U} \subset \mathcal{S}$ is open if and only if $\pi^{-1}(\mathcal{U})$ is open in $\mathcal{M}$. This topology is the strongest topology in which the orbit projection map is continuous, and is characterized by the requirement that π be both a continuous and an open map (i.e., π takes open sets to open sets). In particular, as the continuous image of the connected space $\mathcal{M}$, $\mathcal{S}$ is connected.

For information regarding the topology and geometry of super-space, see Wheeler ([1962], [1964], [1968a,b], [1970]), DeWitt ([1967a,-b, c], [1970]), Ebin [1970], Fischer ([1970], [1983], [1986]), Bour-guignon [1975], Freed and Grossier [1989], and Gil-Medrano and Mi-chor [1991].

We shall need the following information. The action $\Phi : \mathcal{M} \times \mathcal{D} \longrightarrow \mathcal{M}$ is ILH-smooth and proper (see e.g. Ebin [1970]). That the action is proper means that the map

$$\tilde{\Phi} : \mathcal{M} \times \mathcal{D} \longrightarrow \mathcal{M} \times \mathcal{M}; \quad (g, f) \longmapsto (g, f^*g)$$

is a proper mapping, i.e., if $\mathcal{K} \subset \mathcal{M} \times \mathcal{M}$ is compact, then $\tilde{\Phi}^{-1}(\mathcal{K})$ is compact. Equivalently, if $g_n \to \bar{g}$ is a convergent sequence in $\mathcal{M}$ and f_n is a sequence in $\mathcal{D}$ such that $f_n^* g_n \to \tilde{g}$ converges in $\mathcal{M}$, then f_n has a convergent subsequence $f_{k_n} \to \overline{f}$ in $\mathcal{D}$ (and then necessarily $\overline{f}^* \bar{g} = \tilde{g}$). Using the properness of the action, it follows that the orbits $\mathcal{O}_g$ are closed submanifolds in $\mathcal{M}$, that the isotropy groups for the action are compact (see below), and that for each $g \in \mathcal{M}$ there exists a slice S_g for the action Φ, defined as follows:

Definition 6.1 *For the action* $\Phi : \mathcal{M} \times \mathcal{D} \to \mathcal{M}; \quad (g, f) \mapsto f^*g$, *a* slice S_g *at* $g \in \mathcal{M}$ *is a contractible* ILH-*submanifold of* $\mathcal{M}$ *such that*

1. *if* $f \in I_g$, *then* $f^*(S_g) = S_g$ *(i.e., the slice is* I_g*-invariant);*

2. *if* $f \notin I_g$, *then* $f^*(S_g) \cap S_g = \emptyset$ *(equivalently, if* $f \in \mathcal{D}$ *and* $f^*(S_g) \cap S_g \neq \emptyset$, *then* $f \in I_g$*);*

3. *there exists a local cross-section* $\chi : \mathcal{D}/I_g \to \mathcal{D}$ *defined in a neighborhood* $\mathcal{V} \subset \mathcal{D}/I_g$ *of the identity coset such that the map*

$$F : \mathcal{V} \times S_g \longrightarrow \mathcal{M} ; \quad (v, g_1) \longmapsto (\chi(v))^* g_1$$

is an ILH-*diffeomorphism onto an open neighborhood* $\mathcal{U}_g$ *of* $g \in \mathcal{M}$.

From Property (2) of the slice, if $g_1 \in S_g$, then $I_{g_1} \subseteq I_g$, since if $f \in I_{g_1}$, $f^* g_1 = g_1$, so that $f^*(S_g) \cap S_g \neq \emptyset$, and so $f \in I_g$. Thus the isometry groups are locally decreasing (as subgroups) in a slice at g. Together with Property (3) of the slice, this implies that the isometry groups are locally decreasing (as conjugates of subgroups) in an open neighborhood $\mathcal{U}_g$ of g.

As emphasized in Fischer [1970], the key to understanding the structure of superspace is to make the following observation. The isotropy group at $g \in \mathcal{M}$ of the action Φ is given by the isometry group of g, i.e.,

$$\text{Isotropy}_g(\Phi) = \{f \in \mathcal{D} \mid f^* g = g\} = \text{Isometry}_g(M) .$$

Since the isometry groups of different Riemannian metrics on M may have different dimensions and different numbers of components, it follows that in general S is not a manifold. However, it is shown that in all cases $\mathcal{M}/\mathcal{D}$ and $\mathcal{M}/\mathcal{D}_0$ are naturally stratified according to the symmetry type of the geometry. In these stratifications, the strata are manifolds of geometries of the same symmetry type and are organized in such a way that the strata of geometries of higher symmetry are completely contained in the boundary of geometries of lower symmetry (see Fischer [1970] for more details).

In Theorems 6.3 and 7.1 below we show additionally that the differential-topological condition $\deg M = 0$ implies that the stratifications of S, S_0, C, C_0 are orbifolds. In Theorems 8.2 and 8.3 we show that by increasing the topological restrictions on M, S_0 and C_0 are first manifolds and then contractible manifolds, respectively.

Now consider the $\mathcal{D}_0$-restricted action

$$\Phi_0 : \mathcal{M} \times \mathcal{D}_0 \longrightarrow \mathcal{M}; \quad (g, f) \longmapsto f^* g$$

by the normal subgroup $\mathcal{D}_0$ of $\mathcal{D}$. The isotropy for this action at

$g \in \mathcal{M}$ is given by

$$\text{Isotropy}_g(\Phi_0) = I_g \cap \mathcal{D}_0 .$$

Moreover, using the isotropy group $I_g \cap \mathcal{D}_0$ of the Φ_0-action, the slice S_g for the Φ action also works for the Φ_0 action. For example, let S_g be a slice for the action Φ. Since S_g is I_g-invariant, it is also $I_g \cap \mathcal{D}_0$-invariant. Thus (1) of Definition 6.1 holds using the slice S_g and the action Φ_0. Similarly, if $f \in \mathcal{D}_0$ and $f \notin I_g \cap \mathcal{D}_0$, then $f \notin I_g$, and so by (2) of Definition 6.1, $f^*(S_g) \cap S_g = \emptyset$ (equivalently, if $f \in \mathcal{D}_0$ and $f^* S_g \cap S_g \neq \emptyset$, then $f \in I_g$ and thus $f \in I_g \cap \mathcal{D}_0$, the isotropy group of the Φ_0 action). Thus (2) of Definition 6.1 holds using the slice S_g and the Φ_0 action. Since $\mathcal{D}_0$ is the connected component of the identity of $\mathcal{D}$, (3) of Definition 6.1 holds using the slice S_g, the Φ_0 action, and the local cross-section $\mathcal{X}$ restricted to $\mathcal{D}_0/(I_g \cap \mathcal{D}_0)$; namely, $\mathcal{X}_0 : \mathcal{D}_0/(I_g \cap \mathcal{D}_0) \to \mathcal{D}_0$.

We define *quantum superspace* (or $\mathcal{D}_0$-*restricted superspace*) as

$$\mathcal{S}_0 \equiv \mathcal{M}/\mathcal{D}_0$$

and denote points in $\mathcal{S}_0$ by $[g]_0$. Let

$$\pi_0 : \mathcal{M} \longrightarrow \mathcal{M}/\mathcal{D}_0; \quad g \longmapsto [g]_0$$

denote the projection onto the orbit space.

Let $\Gamma = \mathcal{D}/\mathcal{D}_0 = \{[f] = f \circ \mathcal{D}_0 = \mathcal{D}_0 \circ f \mid f \in \mathcal{D}\}$ denote the discrete group of components of $\mathcal{D}$ (for $n = 2$ and M orientable, $\Gamma^+ = \mathcal{D}^+/\mathcal{D}_0$ is the *mapping class group of M*). Now Γ acts on $\mathcal{S}_0 = \mathcal{M}/\mathcal{D}_0$ on the right,

$$\mathcal{S}_0 \times \Gamma \longrightarrow \mathcal{S}_0; \quad ([g]_0, [f]) \longmapsto [f]^*[g]_0 \equiv [f^*g]_0 ,$$

and this action is well-defined since $(f \circ \mathcal{D}_0)^*(\mathcal{D}_0^* g) = \mathcal{D}_0^*(f^*(\mathcal{D}_0^* g)) = \mathcal{D}_0^*((\mathcal{D}_0 \circ f)^* g) = \mathcal{D}_0^*((f \circ \mathcal{D}_0)^* g) = \mathcal{D}_0^*(\mathcal{D}_0^*(f^* g)) = \mathcal{D}_0^*(f^* g)$, so that

$$[(f \circ \mathcal{D}_0)^*(\mathcal{D}_0^* g)]_0 = \pi_0((f \circ \mathcal{D}_0)^*(\mathcal{D}_0^* g)) = \pi_0(\mathcal{D}_0^*(f^* g)) = \pi_0(f^* g) = [f^* g]_0.$$

The orbit space of $\mathcal{S}_0$ by this action is

$$\mathcal{S}_0/\Gamma = \frac{\mathcal{M}/\mathcal{D}_0}{\mathcal{D}/\mathcal{D}_0} \approx \mathcal{M}/\mathcal{D} = \mathcal{S}$$

and the orbit projection map is

$$\pi_{\mathcal{S}} : \mathcal{S}_0 \longrightarrow \mathcal{S} ; \quad [g]_0 \longmapsto [g]$$

(see Section 8 for a further interpretation of this projection map). Thus classical superspace can be studied by using a two-step procedure; first study quantum superspace $\mathcal{S}_0 = \mathcal{M}/\mathcal{D}_0$, and then study $\mathcal{S} = \mathcal{S}_0/\Gamma$. (See also Section 1 for other arguments that $\mathcal{D}_0$-restricted superspace rather than classical superspace is the more appropriate configuration space for the study of canonical quantum gravity.)

Now we consider the notion of orbifolds. In differential geometry, many structures arise which are conveniently described as orbit spaces X/D, where D is a discrete group of diffeomorphisms acting freely and properly discontinuously on a connected n-manifold X. In these circumstances, X/D can be endowed with a differentiable manifold structure such that the orbit projection map $\pi : X \to X/D$ is a differentiable covering. If we relax the condition that D act freely on X, the resulting orbit space X/D is an example of an orbifold. Such spaces were first studied by Satake [1956] as a generalization of the notion of manifold under the name of *V-manifolds*. Thurston [1978] extended the study of these spaces and renamed them *orbifolds*. Thurston's *topological* definition is as follows: *An **orbifold** is a second countable Hausdorff space which is locally homeomorphic to the orbit space of $\mathbf{R}^n$ by the action of a finite group.*

Thus orbifolds can be thought of as generalizations of spaces that globally are quotients of properly discontinuous group actions to spaces that locally have such descriptions. Consequently, orbifolds have at most finite quotient singularities (see Scott [1983] and Davis and Morgan [1984] for more information about orbifolds). The above definition can be extended to include the notion of a *smooth* orbifold by introducing a maximal atlas of orbifold charts and by requiring the overlap maps to be smooth in an orbifold sense. This definition of smooth orbifold generalizes to the ILH setting as follows:

Definition 6.2 *A smooth* ILH*-orbifold* *is a second countable Hausdorff space which is locally homeomorphic to the orbit space of a local* ILH*-manifold by the action of a finite group, and which has an* ILH*-smooth maximal atlas of orbifold charts.*

As an example, suppose that a Lie group $\mathcal{G}$ (perhaps infinite dimensional) acts smoothly and properly on a manifold $\mathcal{X}$ (also perhaps infinite dimensional) such that at each point $x \in \mathcal{X}$ the action has a slice S_x and the isotropy group $\mathcal{G}_x$ at x is finite. Then the orbit space $\mathcal{X}/\mathcal{G}$ of the action is a smooth orbifold. Here the slice S_x is the local manifold, and the orbit space $\mathcal{X}/\mathcal{G}$ at the orbit $[x] \in \mathcal{X}/\mathcal{G}$ is locally modeled on $S_x/\mathcal{G}_x$, a local manifold modulo a finite group. The smoothness of the action then insures that these local models fit together to form a smooth orbifold. Thus for each $x \in \mathcal{X}$ with finite non-trivial isotropy $\mathcal{G}_x \neq \{e\}$, the orbit space $\mathcal{X}/\mathcal{G}$ has a finite quotient singularity at the orbit $[x]$, and this singularity is modeled on $S_x/\mathcal{G}_x$ (see the proof of Theorem 6.3 below for more details on how this works in a specific case).

Now we consider again the action $\Phi : \mathcal{M} \times \mathcal{D} \to \mathcal{M}$. Following the construction in Fischer [1970], we have the following:

Theorem 6.3 (Orbifold structure theorem for $\mathcal{S}$ and $\mathcal{S}_0$) *Let* M *be a compact connected n-manifold, $n \geq 3$. Then $\mathcal{S} \equiv \mathcal{M}/\mathcal{D}$ and $\mathcal{S}_0 \equiv \mathcal{M}/\mathcal{D}_0$ are connected second countable Hausdorff stratified* spaces. *For* $g \in \mathcal{M}$, *let* S_g *be a slice at* g *and let* I_g *be the isometry group of* g. *Then at* $[g] \in \mathcal{S}$, $\mathcal{S}$ *is locally modeled on* S_g/I_g. *Similarly, for* $[g]_0 \in \mathcal{S}_0$, $\mathcal{S}_0$ *is locally modeled on* $S_g/(I_g \cap \mathcal{D}_0)$.

If $\deg M = 0$, *then classical superspace* $\mathcal{S}$ *and quantum superspace* $\mathcal{S}_0$ *are smooth* ILH-*orbifolds.*

Proof: Since $\mathcal{M}$ is connected and since the orbit projection map is continuous, $\mathcal{S}$ is connected. Since the action Φ is proper, $\mathcal{M}/\mathcal{D}$ is a second countable Hausdorff space (see e.g. Fischer [1970]). For each $g \in \mathcal{M}$, the action Φ has a slice S_g (Ebin [1970]), and the isotropy group at each $g \in \mathcal{M}$ is the isometry group I_g of g. From the properties of the slice, if $g_1 \in S_g$, then $I_{g_1} \subseteq I_g$. Thus S_g is I_g-invariant, and so we have the action

$$S_g \times I_g \longrightarrow S_g ; \quad (g, f) \longmapsto f^*g$$

on the slice S_g by the compact group I_g. Let S_g/I_g denote the orbit space for this action, and let $[g]'$ denote points in S_g/I_g. From the properties of the slice, there is a well-defined orbit map

$$g \cdot I_g \longmapsto g \cdot \mathcal{D}$$

($g \cdot I_g = \{f^*g \mid f \in I_g\}$) and thus a well-defined induced map

$$S_g/I_g \longrightarrow \mathcal{M}/\mathcal{D} ; \quad [g]' \longmapsto [g]$$

which from the properties of the slice is a homeomorphism from S_g/I_g onto a neighborhood of $[g] \in \mathcal{M}/\mathcal{D}$. Thus $\mathcal{M}/\mathcal{D}$ is a second countable Hausdorff space modeled on S_g/I_g in a neighborhood of $[g]$.

Similarly, for the action of $\mathcal{D}_0$ on $\mathcal{M}$, the isotropy at $g \in \mathcal{M}$ is given by $I_g \cap \mathcal{D}_0$. With respect to this isotropy (rather than I_g) the slice S_g for the action of $\mathcal{D}$ on $\mathcal{M}$ is also a slice for the action of $\mathcal{D}_0$

on $\mathcal{M}$ since $\mathcal{D}_0$ is the connected component of the identity of $\mathcal{D}$ (see the remarks following Definition 6.1). Thus S_g is $(I_g \cap \mathcal{D}_0)$-invariant, and so we have the action

$$S_g \times (I_g \cap \mathcal{D}_0) \longrightarrow S_g ; \quad (g, f) \longmapsto f^*g$$

on the slice S_g by the compact group $I_g \cap \mathcal{D}_0$. Let $S_g/(I_g \cap \mathcal{D}_0)$ denote the orbit space for this action, and let $[g]_0'$ denote points in $S_g/(I_g \cap \mathcal{D}_0)$. From the properties of the slice, there is a well-defined orbit map

$$g \cdot (I_g \cap \mathcal{D}_0) \longmapsto g \cdot \mathcal{D}_0$$

$(g \cdot (I_g \cap \mathcal{D}_0) = \{f^*g \mid f \in I_g \cap \mathcal{D}_0\})$ and thus a well-defined induced map

$$S_g/(I_g \cap \mathcal{D}_0) \longrightarrow \mathcal{M}/\mathcal{D}_0 ; \quad [g]_0' \longmapsto [g]_0$$

which from the properties of the slice is a homeomorphism from $S_g/(I_g \cap \mathcal{D}_0)$ onto a neighborhood of $[g]_0 \in \mathcal{M}/\mathcal{D}_0$. Thus $\mathcal{M}/\mathcal{D}_0$ is a second countable Hausdorff space modeled on $S_g/(I_g \cap \mathcal{D}_0)$ in a neighborhood of $[g]_0$.

That the spaces $\mathcal{M}/\mathcal{D}$ and $\mathcal{M}/\mathcal{D}_0$ are stratified spaces follows as in Fischer [1970].

If $\deg M = 0$, then for all $g \in \mathcal{M}$, I_g is finite (Proposition 2.2). Since $\mathcal{M}/\mathcal{D}$ is locally modeled on S_g/I_g in a neighborhood of $[g] \in \mathcal{M}/\mathcal{D}$, and since the action is ILH-smooth, it follows that these local models of ILH-manifolds modulo finite groups fit together to form a smooth ILH-orbifold.

Similarly, for all $g \in \mathcal{M}$, $I_g \cap \mathcal{D}_0$ is finite, and since $S_0 = \mathcal{M}/\mathcal{D}_0$ is modeled on $S_g/(I_g \cap \mathcal{D}_0)$ in a neighborhood of $[g]_0$, S_0 is also a smooth ILH-orbifold. ∎

The case of most importance to us is when $n = 3$. In this case, the 3-manifolds with $\deg M = 0$ are given by Theorem 4.3. Thus for M given as in Theorem 4.3, $\mathcal{S}$ and $\mathcal{S}_0$ are smooth ILH-orbifolds.

We shall find a result analogous to Theorem 6.3 for classical and quantum conformal superspace in the next Section.

7   Classical and quantum conformal superspace

In this section we consider classical and quantum conformal superspace (see Fischer-Marsden ([1975], [1977]), Fischer-Tromba ([1984a,-b, c], [1987]), and Fischer-Moncrief ([1994a,b], [1995a,b]), for additional information).

Let M be a compact connected n-manifold. Then the infinite-dimensional abelian group of positive functions $\mathcal{P}$ on M acts on $\mathcal{M}$ by pointwise multiplication,

$$\mathcal{P} \times \mathcal{M} \longrightarrow \mathcal{M}; \quad (p, g) \longmapsto pg$$

For $g \in \mathcal{M}$, let

$$\mathcal{P} \cdot g = \{pg \mid p \in \mathcal{P}\}$$

denote the orbit through g, so that $\mathcal{P} \cdot g$ is the pointwise conformal class of g. The resulting orbit space

$$\mathcal{M}/\mathcal{P} = \{\langle g \rangle \mid g \in \mathcal{M}\}$$

is the space of *pointwise conformal structures on M*. In a natural way $\mathcal{M}/\mathcal{P}$ has the structure of a smooth ILH-manifold, and the projection

$$\mathcal{M} \longrightarrow \mathcal{M}/\mathcal{P}; \quad g \longmapsto \langle g \rangle$$

is an ILH $\mathcal{P}$-principal fiber bundle over $\mathcal{M}/\mathcal{P}$ (see Fischer-Marsden [1977] and Fischer-Tromba [1984a]). Moreover, since $\mathcal{M}$ and $\mathcal{P}$ are contractible and since $\mathcal{P}$ acts freely on $\mathcal{M}$, then, by using the exact homotopy sequence of a fibration and the fact that we are dealing with ILH-manifolds, the resulting orbit space $\mathcal{M}/\mathcal{P}$ is also contractible (see also the proof of Theorem 5.3). Consequently, the $\mathcal{P}$-principal fiber bundle over $\mathcal{M} \to \mathcal{M}/\mathcal{P}$ is bundle isomorphic to the product bundle $(\mathcal{M}/\mathcal{P}) \times \mathcal{P} \to \mathcal{M}/\mathcal{P}$ with projection onto the first factor.

The group $\mathcal{D}$ acts on $\mathcal{M}/\mathcal{P}$ on the right by pull-back,

$$(\mathcal{M}/\mathcal{P}) \times \mathcal{D} \longrightarrow \mathcal{M}/\mathcal{P}; \quad (\langle g \rangle, f) \longmapsto \langle f^* g \rangle$$

For $\langle g \rangle \in \mathcal{M}/\mathcal{P}$, we define $f^* \langle g \rangle = \langle f^* g \rangle$, and let

$$\langle g \rangle \cdot \mathcal{D} = \{ f^* \langle g \rangle \mid f \in \mathcal{D} \}$$

denote the orbit through $\langle g \rangle$. The resulting orbit space

$$C \equiv \frac{\mathcal{M}/\mathcal{P}}{\mathcal{D}} = \{ [\langle g \rangle] \mid \langle g \rangle \in \mathcal{M}/\mathcal{P} \}$$

is the space of *conformal structures on M*, or *(classical) conformal superspace*.

Conformal superspace can be approached in a slightly different manner by considering the semi-direct product $\mathcal{D} \times \mathcal{P}$ of $\mathcal{D}$ with $\mathcal{P}$, namely the product space $\mathcal{D} \times \mathcal{P}$ with semi-direct product group structure

$$(f_1, p_1) \cdot (f_2, p_2) = (f_1 \circ f_2, p_2(f_2^* p_1)) = (f_1 \circ f_2, p_2(p_1 \circ f_2))$$

(see Fischer-Marsden [1977]). Note that this construction is analogous to the construction of the Euclidean group $\boldsymbol{E}(n) = \boldsymbol{O}(n) \times \boldsymbol{R}^n$ as a semi-direct product of the orthogonal group $\boldsymbol{O}(n)$ and the translation group $\boldsymbol{R}^n$ (here $\mathcal{D}$ plays the role of the orthogonal group, and $\mathcal{P}$ plays the role of the translation group). Now $\mathcal{D} \times \mathcal{P}$ acts on $\mathcal{M}$ on the right

$$\mathcal{M} \times (\mathcal{D} \times \mathcal{P}) \longrightarrow \mathcal{M}; \quad (g, (f, p)) \mapsto p f^* g$$

For $g \in \mathcal{M}$, we let $g \cdot (\mathcal{D} \dot{\times} \mathcal{P}) = \{pf^*g \mid p \in \mathcal{P}, f \in \mathcal{D}\}$ denote the $(\mathcal{D} \dot{\times} \mathcal{P})$-orbit through g. Thus $g \cdot (\mathcal{D} \dot{\times} \mathcal{P}) = \{pf^*g \mid p \in \mathcal{P}, f \in \mathcal{D}\}$ is the set of metrics conformally equivalent to g.

The resulting orbit space

$$\frac{\mathcal{M}}{\mathcal{D} \dot{\times} \mathcal{P}} = \{\langle g \rangle \mid g \in \mathcal{M}\} \approx \frac{\mathcal{M}/\mathcal{P}}{\mathcal{D}} \equiv \mathcal{C}$$

is (naturally isomorphic to) conformal superspace $\mathcal{C}$, and gives the construction in a one-step rather than a two-step procedure.

In considering the action of $\mathcal{D}$ on $\mathcal{M}/\mathcal{P}$, recall that for $g \in \mathcal{M}$, $C_g = \{f \in \mathcal{D} \mid f^*g = pg \text{ for some } p \in \mathcal{P}\}$ denotes the *conformal group* of (M, g). Note that $C_g = C_{g'}$ for all g' in the conformal class $\mathcal{P} \cdot g$. Now note that for the action of $\mathcal{D}$ on $\mathcal{M}/\mathcal{P}$, the isotropy group at $\langle g \rangle \in \mathcal{M}/\mathcal{P}$ is given by

$$\text{Isotropy}_{\langle g \rangle} = \{f \in \mathcal{D} \mid f^*\langle g \rangle = \langle g \rangle\} = C_g$$

where the last equality follows from the following sequence of equivalences: $f^*\langle g \rangle = \langle g \rangle \Leftrightarrow \langle f^*g \rangle = \langle g \rangle \Leftrightarrow p_1 f^*g = p_2 g$ for some $p_1, p_2 \in \mathcal{P} \Leftrightarrow f \in C_g$.

In regarding the action of $\mathcal{D} \dot{\times} \mathcal{P}$ on $\mathcal{M}$ (or equivalently, the action $\mathcal{D}$ on $\mathcal{M}/\mathcal{P}$), we have the following information. Firstly, it is shown in Fischer-Marsden [1977] by a direct construction that for each $g \in \mathcal{M}$ there exists a slice $\tilde{S}_g \subset \mathcal{M}$ for this action. However, the existence of a slice does not imply that the action is proper (see below). The isotropy for this action at $g \in \mathcal{M}$ is given by $\text{Isotropy}_g \equiv \tilde{C}_g \equiv \{(f, p) \mid pf^*g = g\}$. Noting that projection onto the first factor $\pi_1 : \tilde{C}_g \to C_g$; $(f, p) \mapsto f$ is a group isomorphism, we have $\tilde{C}_g \approx C_g$. Thus the isotropy groups of the actions of $\mathcal{D}$ on $\mathcal{M}/\mathcal{P}$ and of $\mathcal{D} \dot{\times} \mathcal{P}$ on $\mathcal{M}$ are isomorphic.

In general, however, these actions need not be proper, since a proper action must have compact isotropy groups, and since in general $\tilde{C}_g \approx C_g$ need not be compact. For example, if (M, g) is conformally equivalent to a standard n-sphere, then C_g is not compact. However, a theorem of Lelong-Ferrand ([1969], [1971]) and Obata [1971] states that this is the only case in which C_g is non-compact, i.e. *if (M, g) is a compact connected Riemannian n-manifold, $n \geq 3$, then the conformal group C_g is non-compact if and only if (M, g) is conformally equivalent to the standard sphere.* In particular, if M is not diffeomorphic to S^n, then (M, g) is not conformally equivalent to a standard sphere, so that in this case C_g is compact for every $g \in \mathcal{M}$. Furthermore, using the methods in Itoh [1991], one can prove that for $M \not\approx S^n$, the action of $\mathcal{D} \times \mathcal{P}$ on $\mathcal{M}$ and the action of $\mathcal{D}$ on $\mathcal{M}/\mathcal{P}$ are proper actions.

As with superspace, we can also consider the $\mathcal{D}_0$-restricted actions

$$\mathcal{M} \times (\mathcal{D}_0 \times \mathcal{P}) \longrightarrow \mathcal{M}; \quad (g, (f, p)) \mapsto pf^*g$$

and

$$(\mathcal{M}/\mathcal{P}) \times \mathcal{D}_0 \longrightarrow \mathcal{M}/\mathcal{P}; \quad (\langle g \rangle, f) \longmapsto \langle f^*g \rangle .$$

Note that since $\mathcal{P}$ is contractible, the connected component of the identity of $\mathcal{D} \times \mathcal{P}$ is $(\mathcal{D} \times \mathcal{P})_0 = \mathcal{D}_0 \times \mathcal{P}$.

As with quantum superspace, we refer to the resulting orbit space

$$C_0 \equiv \frac{\mathcal{M}/\mathcal{P}}{\mathcal{D}_0} \approx \frac{\mathcal{M}}{\mathcal{D}_0 \times \mathcal{P}}$$

as quantum conformal superspace (or $\mathcal{D}_0$-restricted conformal superspace). For $g \in \mathcal{M}$, let $[\langle g \rangle]_0 \in \frac{\mathcal{M}/\mathcal{P}}{\mathcal{D}_0} = C_0$ denote the orbit in C_0 and let $(g)_0 \in \frac{\mathcal{M}}{\mathcal{D}_0 \times \mathcal{P}}$ denote the orbit in $\frac{\mathcal{M}}{\mathcal{D}_0 \times \mathcal{P}}$.

In studying the structure of C and C_0, it is usually more convenient to study the action of $\mathcal{D} \times \mathcal{P}$ and $\mathcal{D}_0 \times \mathcal{P}$ on $\mathcal{M}$ rather than the action of $\mathcal{D}$ and $\mathcal{D}_0$ on $\mathcal{M}/\mathcal{P}$. This is because $\mathcal{D} \times \mathcal{P}$ and $\mathcal{D}_0 \times \mathcal{P}$ act directly on $\mathcal{M}$ whereas $\mathcal{D}$ and $\mathcal{D}_0$ act on the orbit space $\mathcal{M}/\mathcal{P}$.

Now, analogous to Theorem 6.3, we have the following:

Theorem 7.1 (Orbifold structure theorem for C and C_0) *Let M be a compact connected n-manifold, $n \geq 3$, $M \not\approx S^n$. Then classical conformal superspace $C \equiv \frac{M/P}{D} \approx \frac{M}{D \dot\times P}$ and quantum conformal superspace $C_0 \equiv \frac{M/P}{D_0} \approx \frac{M}{D_0 \dot\times P}$ are connected second countable Hausdorff stratified spaces. For $g \in M$, let $\tilde{S}_g$ be a slice at g with respect to the action of $D \dot\times P$ on M and let $\tilde{C}_g$ be the isotropy group at g. Then at $(g) \in C$, C is modeled on $\tilde{S}_g/\tilde{C}_g$. Similarly, for $(g)_0 \in C_0$, C_0 is modeled on $\tilde{S}_g/(\tilde{C}_g \cap (D_0 \dot\times P))$.*

If $\deg M = 0$, then conformal superspace C and quantum conformal superspace C_0 are smooth ILH-orbifolds.

Proof: As the continuous images of the connected set M/P, C and C_0 are connected. Since $M \not\approx S^n$, using the methods in Itoh [1991], one can show that the actions of $D \dot\times P$ and $D_0 \dot\times P$ on M are proper. From this it follows that the resulting orbit spaces are second countable Hausdorff spaces, and that there exist slices for these actions. We shall call such slices conformal slices. For $g \in M$, let $\tilde{S}_g \subset M$ denote a conformal slice at g for this action, and let $\tilde{C}_g$ denote the isotropy group at g. Since $\tilde{S}_g$ is $\tilde{C}_g$-invariant, we have the action

$$\tilde{S}_g \times \tilde{C}_g \longrightarrow \tilde{S}_g \; ; \quad (g, (f, p)) \longmapsto pf^*g$$

on the slice $\tilde{S}_g$ by the compact group $\tilde{C}_g$. Let $\tilde{S}_g/\tilde{C}_g$ denote the orbit space for this action, and let $(g)'$ denote points in $\tilde{S}_g/\tilde{C}_g$. From the properties of the slice, if $g_1 \in \tilde{S}_g$, then $\tilde{C}_{g_1} \subseteq \tilde{C}_g$. Thus there is a well-defined orbit map

$$g \cdot \tilde{C}_g \longmapsto g \cdot (D \dot\times P)$$

$(g \cdot \tilde{C}_g = \{pf^*g \mid (f, p) \in \tilde{C}_g\})$ and thus a well-defined induced map

$$\tilde{S}_g/\tilde{C}_g \longrightarrow M/(D \dot\times P) \; ; \quad (g)' \longmapsto (g)$$

which from the properties of the slice is a homeomorphism from $\tilde{S}_g/\tilde{C}_g$ onto a neighborhood of $(g) \in M/(D \dot\times P)$. Thus $M/(D \dot\times P)$

is a second countable Hausdorff space modeled on $\tilde{S}_g/\tilde{C}_g$ in a neighborhood of (g).

Similarly, as in the proof of Theorem 6.3, the isotropy for the action of $\mathcal{D}_0 \dot\times \mathcal{P}$ on $\mathcal{M}$ is given by $\tilde{C}_g \cap (\mathcal{D}_0 \dot\times \mathcal{P})$, and the slice $\tilde{S}_g$ for the action of $\mathcal{D} \dot\times \mathcal{P}$ on $\mathcal{M}$ is also a slice for the action of $\mathcal{D}_0 \dot\times \mathcal{P}$ on $\mathcal{M}$. Thus there is a homeomorphism from $\tilde{S}_g/(\tilde{C}_g \cap (\mathcal{D}_0 \dot\times \mathcal{P}))$ onto a neighborhood of $(g)_0 \in \mathcal{M}/(\mathcal{D}_0 \dot\times \mathcal{P})$, and thus $\mathcal{M}/(\mathcal{D}_0 \dot\times \mathcal{P})$ is modeled on $\tilde{S}_g/(\tilde{C}_g \cap (\mathcal{D}_0 \dot\times \mathcal{P}))$ in a neighborhood of $(g)_0$.

That the spaces $\mathcal{M}/(\mathcal{D} \dot\times \mathcal{P})$ and $\mathcal{M}/(\mathcal{D}_0 \dot\times \mathcal{P})$ are stratified spaces then follows as in Fischer [1970] and Fischer-Marsden [1977].

Now suppose $\deg M = 0$ so that $M \not\approx S^n$. Another theorem of Lelong-Ferrand ([1969], [1971]) and Obata [1971] states that if (M, g) is a compact connected Riemannian n-manifold, $n \geq 3$, then the identity component C_g^0 of C_g is a subgroup of the isometry group I_{pg} of some conformally related metric pg (for some $p \in \mathcal{P}$) if and only if (M, g) is not conformally equivalent to the standard sphere. Thus if $\deg M = 0$, M cannot be diffeomorphic to an n-sphere, and (M, g) cannot be conformally equivalent to a standard n-sphere. Thus, firstly, for every $g \in \mathcal{M}$, C_g is compact, and then secondly, $C_g^0 \subseteq I_{pg}$ for some $p \in \mathcal{P}$, and hence $C_g^0 \subseteq I_{pg}^0$. But since $\deg M = 0$, I_{pg} is a finite group, and thus $C_g^0 = I_{pg}^0 = \{\mathrm{id}_M\}$. Thus C_g is discrete and compact, and hence finite. Thus any metric on a compact $\deg M = 0$ manifold must have a finite conformal group (as well as a finite isometry group). Since the isotropy groups of the ILH-smooth action of $\mathcal{D} \dot\times \mathcal{P}$ on $\mathcal{M}$ are isomorphic to the finite conformal groups, the orbit space $\mathcal{M}/(\mathcal{D}_0 \dot\times \mathcal{P})$ is modeled on $\tilde{S}_g/\tilde{C}_g \approx$ ILH-manifold modulo a finite group. Moreover, since the action of $\mathcal{D} \dot\times \mathcal{P}$ on $\mathcal{M}$ is ILH-smooth, it follows that these local models fit together in an ILH-smooth way. Consequently, conformal superspace $\mathcal{C}$ is a smooth ILH-orbifold.

Similarly, noting that $\tilde{C}_g \cap (\mathcal{D}_0 \dot{\times} \mathcal{P}) \approx C_g \cap \mathcal{D}_0$ is a finite group if $\deg M = 0$, then arguing as above, we see that $\mathcal{C}_0$ is a smooth ILH-orbifold modeled on $\tilde{S}_g/(\tilde{C}_g \cap (\mathcal{D}_0 \dot{\times} \mathcal{P}))$ in a neighborhood of $(g)_0 \in \mathcal{M}/(\mathcal{D}_0 \dot{\times} \mathcal{P})$. ∎

8 The structure of quantum superspace and quantum conformal superspace

In Sections 6 and 7 we have seen that when $\deg M = 0$ superspace, conformal superspace, and their quantum analogs are orbifolds. In this Section we are interested in a more detailed analysis of the structure of quantum superspace and quantum conformal superspace. An important question regarding this structure is under what circumstances "orbifold-type" singularities do not occur. The case of most interest to us is for $n = 3$. Thus we wish to focus our attention on those 3-manifolds M for which $\mathcal{D}_0$ acts freely on $\mathcal{M}$ and on $\mathcal{M}/\mathcal{P}$. Clearly, a necessary condition for $\mathcal{D}_0$ to act freely on $\mathcal{M}$ is that $\deg M = 0$. Such 3-manifolds are given by Theorem 4.3. Thus we are interested in the subclass of 3-manifolds with $\deg M = 0$ for which $\mathcal{D}_0$ acts freely on $\mathcal{M}$. For this same subclass, $\mathcal{D}_0$ will then act freely on $\mathcal{M}/\mathcal{P}$; i.e., no non-trivial isometries implies no non-trivial conformal isometries (see the Proof of Theorem 8.2). Now we introduce the following:

Definition 8.1 *Let M be a compact connected orientable 3-manifold. Then M is **admissible** if either*

(1) M is an irreducible $K(\pi, 1)$-manifold with $\deg M = 0$;

(2) M is a composite manifold (i.e., M is a connected sum of two or more prime manifolds), diffeomorphic to a connected sum of the

form

$$M \approx \underbrace{S^3/\Gamma_1 \# \ldots \# S^3/\Gamma_k}_{\text{spherical-factors}} \# \underbrace{(S^1 \times S^2)_1 \# \ldots \# (S^1 \times S^2)_l}_{\text{handles (or wormholes)}}$$

$$\# \underbrace{K(\pi_1, 1) \# \ldots \# K(\pi_m, 1)}_{K(\pi,1)\text{-factors}}$$

where (1), (2), and (3) of Theorem 2.11 hold, where $k + l + m \geq 2$ (since M is a composite manifold), and where either $m \geq 1$, or if $m = 0$, then $k \geq 1$ and at least one of the spherical-type factors S^3/Γ_i is not a lens space.

Note that from Theorem 4.3 and the remarks following, in either case (1) or (2), deg $M = 0$. Also note that (1) can be refined to

(1') M is a $K(\pi, 1)$-manifold of flat type F_6, or M is an irreducible $K(\pi, 1)$-manifold of non-flat

   type with deg $M = 0$.

Comparing the definition of admissible with Theorem 4.3 when the Poincaré conjecture is true, we see that if deg $M = 0$, then M is admissible, with the possible exception of a manifold of the form S^3/Γ that is not diffeomorphic to a spherical space form; such a manifold automatically has degree $= 0$. Note, however, that such exceptions do not exist if the conjecture that *every finite group acting freely on S^3 is equivalent to a standard orthogonal action of a finite subgroup of $SO(4)$ acting on S^3* is true; see the remarks following Theorem 2.11. Thus if both the Poincaré conjecture and this latter conjecture are true, then M is admissible if and only if deg $M = 0$.

Lastly, we remark that if M is admissible, then M satisfies the hypotheses of Theorem 3.3. (i.e. being admissible is more restrictive than satisfying the hypotheses of Theorem 3.3). Now we have the following:

Theorem 8.2 (Manifold structure theorem) *Let M be a compact connected orientable admissible 3-manifold. Then the following actions are free and proper*

$$\mathcal{M} \times \mathcal{D}_0 \longrightarrow \mathcal{M} \; ; \quad (g, f) \longmapsto f^*g$$

$$(\mathcal{M}/\mathcal{P}) \times \mathcal{D}_0 \longrightarrow \mathcal{M}/\mathcal{P} \; ; \quad (\langle g \rangle, f) \longmapsto \langle f^*g \rangle$$

$$\mathcal{M} \times (\mathcal{D}_0 \dot{\times} \mathcal{P}) \longrightarrow \mathcal{M} \; ; \quad (g, (f, p)) \longmapsto p f^*g$$

Consequently, (1) the resulting orbit spaces $\mathcal{S}_0 = \mathcal{M}/\mathcal{D}_0$ and $\mathcal{C}_0 = \frac{\mathcal{M}/\mathcal{P}}{\mathcal{D}_0} \approx \frac{\mathcal{M}}{\mathcal{D}_0 \dot{\times} \mathcal{P}}$ are connected and simply connected ILH-*manifolds, (2) the orbit projection maps*

$$\mathcal{M} \longrightarrow \mathcal{M}/\mathcal{D}_0$$

and

$$\mathcal{M}/\mathcal{P} \longrightarrow \frac{\mathcal{M}/\mathcal{P}}{\mathcal{D}_0}$$

are ILH $\mathcal{D}_0$-*principal fiber bundles with base spaces $\mathcal{M}/\mathcal{D}_0$ and $\frac{\mathcal{M}/\mathcal{P}}{\mathcal{D}_0}$, respectively, and (3) the orbit projection map*

$$\mathcal{M} \longrightarrow \frac{\mathcal{M}}{\mathcal{D}_0 \dot{\times} \mathcal{P}} \approx \frac{\mathcal{M}/\mathcal{P}}{\mathcal{D}_0}$$

is an ILH $(\mathcal{D}_0 \dot{\times} \mathcal{P})$-*principal fiber bundle with base space $\frac{\mathcal{M}}{\mathcal{D}_0 \dot{\times} \mathcal{P}} \approx \frac{\mathcal{M}/\mathcal{P}}{\mathcal{D}_0}$.*

Moreover, $\mathcal{S}_0$ and $\mathcal{C}_0$ are homotopy equivalent, with homotopy groups given by

$$\pi_{i+1}(\mathcal{S}_0) \approx \pi_{i+1}(\mathcal{C}_0) \approx \pi_i(\mathcal{D}_0) \; ; \quad i \geq 0 \,.$$

Proof: First consider the right action of $\mathcal{D}_0$ on $\mathcal{M}$,

$$\mathcal{M} \times \mathcal{D}_0 \longrightarrow \mathcal{M} \,.$$

For this action, the isotropy group for $g \in \mathcal{M}$ is given by

$$\text{Isotropy}_g = \{f \in \mathcal{D}_0 \mid f^* g = g\} = I_g \cap \mathcal{D}_0 \,.$$

Since M is compact and $\deg M = 0$, I_g is a finite group (Proposition 2.2). Hence $\text{Isotropy}_g = I_g \cap \mathcal{D}_0$ is a finite group. Let $f \in \text{Isotropy}_g = I_g \cap \mathcal{D}_0$ and assume $f \neq \text{id}_M$. Since $I_g \cap \mathcal{D}_0$ is a finite group, f must have finite order, so f is a non-trivial symmetry of M. Since f is in $\mathcal{D}_0$, f is isotopic (i.e. homotopic through a curve of diffeomorphisms) and hence homotopic to the identity. Since M is admissible, both Theorem 3.3 applies and $\deg M = 0$. Thus f is toral and so embeds in an effective action of $\boldsymbol{SO}(2)$ on M, thereby contradicting $\deg M = 0$. Thus $f = \text{id}_M$, and $\mathcal{D}_0$ acts freely on $\mathcal{M}$.

Now $\mathcal{D}_0$ acts freely and properly on $\mathcal{M}$, and so $\mathcal{M}/\mathcal{D}_0$ is an ILH-manifold and the orbit projection map $\mathcal{M} \longrightarrow \mathcal{M}/\mathcal{D}_0$ has the structure of a $\mathcal{D}_0$-principal fiber bundle over $\mathcal{S}_0$.

Now consider the action of $\mathcal{D}_0$ on $\mathcal{M}/\mathcal{P}$,

$$(\mathcal{M}/\mathcal{P}) \times \mathcal{D}_0 \longrightarrow \mathcal{M}/\mathcal{P} \,.$$

The isotropy for this action at $\langle g \rangle \in \mathcal{M}/\mathcal{P}$ is given by

$$\text{Isotropy}_{\langle g \rangle} = \{f \in \mathcal{D}_0 \mid f^* \langle g \rangle = \langle g \rangle\} = C_g \cap \mathcal{D}_0 \,.$$

As in the proof of Theorem 7.1, since M is compact and $\deg M = 0$, C_g is a finite group. Thus $\text{Isotropy}_{\langle g \rangle} = C_g \cap \mathcal{D}_0$ is a finite group. Thus if $f \in C_g \cap \mathcal{D}_0$ is non-trivial, then f is a non-trivial symmetry homotopic to the identity, hence toral, and thus contradicting $\deg M = 0$. Thus $f = \{\text{id}_M\}$ and the action $\mathcal{D}_0$ on $\mathcal{M}/\mathcal{P}$ is free. Since $\deg M = 0$, $M \not\approx S^3$, and so this action is also proper (see the proof of Theorem 7.1). Thus $\mathcal{D}_0$ acts freely and properly on $\mathcal{M}/\mathcal{P}$, and thus the orbit space $\frac{\mathcal{M}/\mathcal{P}}{\mathcal{D}_0}$ is an ILH-manifold, and the orbit projection map $\mathcal{M}/\mathcal{P} \longrightarrow \frac{\mathcal{M}/\mathcal{P}}{\mathcal{D}_0}$ has the structure of a $\mathcal{D}_0$-principal fiber bundle over $\mathcal{C}_0$.

For the action of $\mathcal{D}_0 \dot{\times} \mathcal{P}$ on $\mathcal{M}$, the isotropy group at $g \in \mathcal{M}$ is given by $\tilde{C}_g \cap (\mathcal{D}_0 \dot{\times} \mathcal{P}) \approx C_g \cap \mathcal{D}_0$, a finite group. The result now follows as above.

That $\mathcal{S}_0 = \mathcal{M}/\mathcal{D}_0$ and $\mathcal{C}_0 = \frac{\mathcal{M}/\mathcal{P}}{\mathcal{D}_0}$ are homotopy equivalent follows from the fact that $\mathcal{P}$ is contractible. Moreover, since $\mathcal{M} \to \mathcal{S}_0$ is a fibration, it follows from the exact homotopy sequence for a fibration (see e.g. Hu [1959]) that

$$\pi_{i+1}(\mathcal{S}_0) \approx \pi_{i+1}(\mathcal{C}_0) \approx \pi_i(\mathcal{D}_0) ; \quad i \geq 0 .$$

In particular, since $\mathcal{D}_0$ is connected, $\mathcal{S}_0$ and $\mathcal{C}_0$ are simply connected. ∎

We remark that since $\mathcal{M}$ and $\mathcal{M}/\mathcal{P}$ are contractible manifolds, the $\mathcal{D}_0$-principal fiber bundles of Theorem 5.1 $\mathcal{M} \to \mathcal{S}_0$ and $\mathcal{M}/\mathcal{P} \to \mathcal{C}_0$ are universal $\mathcal{D}_0$-principal fiber bundles (see Milnor [1956a,b]) with total space $E_{\mathcal{D}_0} = \mathcal{M}$ (or $\mathcal{M}/\mathcal{P}$) and base space $B_{\mathcal{D}_0} = \mathcal{S}_0$ (or $\mathcal{C}_0$). Using either universal bundle, the isomorphism classes of $\mathcal{D}_0$-principal fiber bundles over any second countable Hausdorff space $\mathcal{X}$ are then in bijective correspondence with the homotopy classes of mappings $[\mathcal{X}, \mathcal{S}_0]$ (or $[\mathcal{X}, \mathcal{C}_0]$).

Let $\Gamma = \mathcal{D}/\mathcal{D}_0$ denote the group of components of $\mathcal{D}$ (see also Section 6). Then Γ acts on $\mathcal{S}_0$,

$$\mathcal{S}_0 \times \Gamma \longrightarrow \mathcal{M}/\mathcal{D}_0 ; \quad ([g]_0, [f]) \longmapsto [f^*g]_0 .$$

The resulting orbit space is given by

$$\mathcal{S}_0/\Gamma = \frac{\mathcal{M}/\mathcal{D}_0}{\mathcal{D}/\mathcal{D}_0} \approx \mathcal{M}/\mathcal{D} = \mathcal{S} ,$$

with orbit projection map

$$\mathcal{M}/\mathcal{D}_0 \longrightarrow \mathcal{M}/\mathcal{D} ; \quad [g]_0 \longmapsto [g] .$$

Now assume that M is admissible so that $\deg M = 0$. Then from Theorem 6.3, S is a connected ILH-orbifold, and from Theorem 8.2, S_0 is a connected simply connected ILH-manifold. Thus we can interpret this state of affairs by saying that S_0 is the universal orbifold covering of S, where the ILH-orbifold S_0 actually turns out to be an ILH-manifold.

Similarly, we have the projection

$$C_0 = \frac{M/P}{D_0} \longrightarrow C = \frac{M/P}{D} \; ; \quad [\langle g \rangle]_0 \longmapsto [\langle g \rangle] \;.$$

Then, if M is admissible, C_0 is a simply connected and hence universal ILH-orbifold covering of the ILH-orbifold C, which again turns out to be an ILH-manifold covering an ILH-orbifold.

We remark that if $\deg M = 0$ but if M is not admissible, then D_0 may not act freely on M and on M/P. In this case, we still have the projections $S_0 \to S$ and $C_0 \to C$, which are still projections of smooth ILH-orbifolds. However, if D_0 does not act freely on M, the homotopy of the orbit space M/D_0 cannot be computed using the exact homotopy sequence of a fibration, and indeed, M/D_0 need not be simply connected. In this case, the projection $S_0 \to S$ would not be a universal covering projection. Similarly, the orbifold projection $C_0 \to C$ would not be universal.

Increasing the topological restrictions on M now gives the following stronger result:

Theorem 8.3 (Contractible Manifold Theorem) *Let M be a Haken 3-manifold with $\deg M = 0$. Then D_0 and $D_0 \times P$ are contractible groups. From Theorem 8.2, D_0 acts freely and properly on M and on M/P, and $D_0 \times P$ acts freely and properly on M. Consequently, $S_0 = M/D_0$ and $C_0 = \frac{M/P}{D_0} \approx \frac{M}{D_0 \times P}$ are contractible ILH-manifolds. Therefore, the orbit projection maps*

$$M \longrightarrow M/D_0$$

and

$$\mathcal{M}/\mathcal{P} \longrightarrow \frac{\mathcal{M}/\mathcal{P}}{\mathcal{D}_0}$$

are ILH $\mathcal{D}_0$*-principal fiber bundles with contractible base spaces* $\mathcal{M}/\mathcal{D}_0$ *and* $\frac{\mathcal{M}/\mathcal{P}}{\mathcal{D}_0}$ *respectively, and thus are principal fiber bundle isomorphic to the product bundles* $\mathcal{S}_0 \times \mathcal{D}_0 \to \mathcal{S}_0$ *and* $\mathcal{C}_0 \times \mathcal{D}_0 \to \mathcal{C}_0$, *respectively, with projection onto the first factors.*

Similarly, the orbit projection map

$$\mathcal{M} \longrightarrow \frac{\mathcal{M}}{\mathcal{D}_0 \overset{.}{\times} \mathcal{P}}$$

is an ILH $(\mathcal{D}_0 \overset{.}{\times} \mathcal{P})$*-principal fiber bundle with contractible base space* $\frac{\mathcal{M}}{\mathcal{D}_0 \overset{.}{\times} \mathcal{P}} \approx \frac{\mathcal{M}/\mathcal{P}}{\mathcal{D}_0}$ *and thus is principal fiber bundle isomorphic to the product bundle* $\mathcal{C}_0 \times (\mathcal{D}_0 \overset{.}{\times} \mathcal{P}) \to \mathcal{C}_0$ *with projection onto the first factor.*

Proof: Since M is a Haken manifold, it is an irreducible $K(\pi, 1)$-manifold, and since $\deg M = 0$, M is admissible. Thus from Theorem 8.2, the three actions described above are free and proper, so that the resulting orbit spaces $\mathcal{S}_0 = \mathcal{M}/\mathcal{D}_0$ and $\mathcal{C}_0 = \frac{\mathcal{M}/\mathcal{P}}{\mathcal{D}_0} \approx \frac{\mathcal{M}}{\mathcal{D}_0 \overset{.}{\times} \mathcal{P}}$ are ILH-manifolds. Since M is a Haken manifold with $\deg M = 0$, from Theorem 5.3, $\mathcal{D}_0$ is contractible. From Theorem 8.2, all of the homotopy groups of $\mathcal{S}_0$ and $\mathcal{C}_0$ vanish, and thus these spaces are weakly homotopy equivalent to a point. Since these spaces are ILH-manifolds, they are thus contractible (Palais [1966]). The resulting principal fiber bundles are then isomorphic to the respective product bundles $\mathcal{S}_0 \times \mathcal{D}_0$, $\mathcal{C}_0 \times \mathcal{D}_0$, and $\mathcal{C}_0 \times (\mathcal{D}_0 \overset{.}{\times} \mathcal{P})$ with projection onto their first factors. ∎

Thus for Haken 3-manifolds with $\deg M = 0$, the $\mathcal{D}_0$-principal fiber bundles $\mathcal{M} \longrightarrow \mathcal{S}_0$ and $\mathcal{M}/\mathcal{P} \longrightarrow \mathcal{C}_0$ are bundle isomorphic to the trivial product bundles $\mathcal{S}_0 \times \mathcal{D}_0 \to \mathcal{S}_0$ and $\mathcal{C}_0 \times \mathcal{D}_0 \to \mathcal{C}_0$, respectively. In particular, any of these bundles possess global cross-sections. The equivalent physical statement is that a continuous choice of gauge exists for these configuration spaces, and so a Gribov ambiguity does not occur for these spaces (Gribov [1978], Singer [1978], Mitter and Viallet [1981]).

Note the analogy of Theorem 8.3 to Teichmüller's theorem, suitably formulated (see e.g. Earle-Eells [1969], Fischer-Tromba ([1984a, b, c], [1987]), and Tromba [1992]):

Theorem 8.4 *Let S be a compact connected orientable 2-manifold such that $\deg S = 0$ (so that S is a compact connected orientable surface with $\mathrm{genus}(S) = k \geq 2$). Let $\mathcal{T}_k(S)$ denote the Teichmüller space of S.*

Then $S_0 = \mathcal{M}/\mathcal{D}_0$ is a contractible ILH-*manifold and the Teichmüller space $\mathcal{T}_k(S) \equiv \mathcal{C}_0 = \frac{\mathcal{M}/\mathcal{P}}{\mathcal{D}_0}$ is a finite-dimensional contractible manifold (in fact a cell of dimension $6k - 6$).*

This analogy adds another point of similarity between Haken manifolds and surfaces $\not\approx S^2$; see e.g. Freedman and Yau [1983]. Note that for a compact connected orientable surface S with genus $= k$, $\deg S = 0 \Leftrightarrow k \geq 2 \Leftrightarrow S$ is a non-flat $K(\pi, 1)$-manifold $\Leftrightarrow S \not\approx S^2$, T^2. Then, under any of these cases, S_0 and $\mathcal{C}_0$ are contractible manifolds.

Note also, however, the dissimilarity between the $n = 3$ (Theorem 8.3) and $n = 2$ (Theorem 8.4) cases. In Theorem 8.3, M is a Haken and hence a prime manifold, whereas in Theorem 8.4, genus $(S) \geq 2$, and hence S is not prime.

Summarizing Theorems 8.2 and 8.3, for a compact connected orientable admissible 3-manifold, $\mathcal{D}_0$ and $\mathcal{D}_0 \times \mathcal{P}$ act freely and properly on $\mathcal{M}$. Consequently, S_0 and $\mathcal{C}_0$ are connected and simply connected ILH-manifolds. If M is a Haken 3-manifold with $\deg M = 0$, then $\mathcal{D}_0$ is contractible, and hence $\mathcal{S}_0$ and $\mathcal{C}_0$ are contractible ILH-manifolds.

Now if M is a Haken manifold, then, in particular, M is an irreducible $K(\pi, 1)$-manifold. It is reasonable to conjecture that the Haken condition is not critical, and that $\mathcal{D}_0$, $\mathcal{S}_0$, and $\mathcal{C}_0$ are contractible under the more general condition that M be an irreducible $K(\pi, 1)$-manifold with $\deg M = 0$ (see also the remarks following Theorem 5.3).

We remark that if $\mathcal{D}_0$ is not contractible, then from Theorem 8.2 $\mathcal{S}_0$ and $\mathcal{C}_0$ also have non-vanishing homotopy and thus are also not contractible. Methods to study the structure of $\mathcal{D}_0$ appear in Giulini ([1994], [1995]) and the references cited therein. Using these methods it is apparent that $\mathcal{D}_0$ is not contractible unless M is a $K(\pi, 1)$-manifold. Thus if the aforementioned conjecture is true, then $\mathcal{S}_0$ and $\mathcal{C}_0$ would be contractible ILH-manifolds if and only if M is an irreducible $K(\pi, 1)$-manifold with $\deg M = 0$.

Finally, we remark that questions involving the structure of $\mathcal{C}$ and $\mathcal{C}_0$ appear in the problem of reduction, i.e., in the problem of reducing the general Hamiltonian formulation of general relativity with constraints to an unconstrained Hamiltonian system on some reduced phase space in which the reduced Hamiltonian would not be degenerate. Since the most common approach to reduction uses conformal methods, $\mathcal{C}$ and $\mathcal{C}_0$ are natural candidates for reduced configuration spaces. The success of such a reduction program would have two major benefits. On the one hand, there would be no arbitrary lapse function and shift vector field that must be specified in advance before the evolution of the dynamical variables could take place. On the other hand, the reduced canonical variables would be free, or unconstrained, and so the dynamical variables would carry complete information about the true degrees of freedom of the gravitational field (see Fischer-Moncrief ([1994a,b], [1995a,b]) for further information). Lastly, we remark that the structure of C_0 will surely play a role in any theory of quantum gravity based on reduced configuration spaces and conformal methods.

References

[1] Arnowitt, R, Deser, S, and Misner, C (1962), *The dynamics of general relativity*, in *Gravitation: an introduction to current research*, L. Witten, editor, John Wiley and Sons, Inc., New York.

[2] Balachandran, A P (1989), *Classical topology and quantum phases: quantum mechanics*, in *Geometrical and Algebraic Aspects of Nonlinear Field Theory*, S. De Filippo, M. Marinaro, G. Marmo, and G. Vilasi, editors, Elsevier Science Publishers.

[3] Bourguignon, J-P (1975), *Une stratification de l'espace des structures riemanniennes*, Compositio Math. **30**, 1-41.

[4] Bröcker, T, and Jänich, K (1982), *Introduction to Differential Topology*, Cambridge University Press, Cambridge.

[5] Davis, J F, and Milgram, R J (1985), *A Survey of the Spherical Space Form Problem*, Mathematical Reports, Volume 2, Part 2, Harwood Academic Publishers, New York.

[6] Davis, M, and Morgan, J (1984), *Finite group actions on homotopy 3-spheres*, in *The Smith Conjecture*, Morgan, J, and Bass, H, editors, Academic Press, Inc., New York.

[7] DeWitt, B (1967a), *Quantum theory of gravity. I. The canonical theory*, Physical Review **160**, 1113-1148.

[8] DeWitt, B (1967b), *Quantum theory of gravity, II. The manifestly covariant theory*, Physical Review **162**, 1195-1239.

[9] DeWitt, B (1967c), *Quantum theory of gravity, III. Application of the covariant theory*, Physical Review **162**, 1239-1256.

[10] DeWitt, B (1970), *Spacetime as a sheaf of geodesics in super-space*, in *Relativity*, M. Carmeli, S. Fickler, and L. Witten, editors, Plenum Press, New York.

[11] Earle, C, and Eells, J (1969), *A fibre bundle description of Teichmüller theory*, J. Diff. Geom., **3**, 19-43.

[12] Ebin, D (1970), *The space of Riemannian metrics*, Proc. Symp. Pure Math., Amer. Math. Soc. **15**, 11-40.

[13] Ebin, D, and Marsden, J (1970), *Groups of diffeomorphisms and the motion of an incompressible fluid*, Annals of Mathematics **92**, 102-163.

[14] Edmonds, A (1985), *Transformation groups and low-dimensional manifolds*, in *Group Actions on Manifolds*, Contemporary Mathematics, Volume 36, R. Schultz, editor, 339-366.

[15] Ellis, G (1971), *Topology and cosmology*, General Relativity and Gravitation **2**, 7-21.

[16] Fischer, A (1970), *The theory of superspace*, in *Relativity*, M. Carmeli, S. Fickler, and L. Witten, editors, Plenum Press, New York.

[17] Fischer, A, and Marsden, J (1975), *Deformations of the scalar curvature*, Duke Mathematical Journal **42**, 519-547.

[18] Fischer, A, and Marsden, J (1977), *The manifold of conformally equivalent metrics*, Canadian Journal of Mathematics, **1**, 193-209.

[19] Fischer, A, and Moncrief, V (1994a), *Reducing Einstein's equations to an unconstrained Hamiltonian system on the cotangent bundle of Teichmüller space*, in *Physics on Manifolds, Pro-*

ceedings on the International Colloquium in honour of Yvonne Choquet-Bruhat, M. Flato, R. Kerner, and A. Lichnerowicz, editors, Kluwer Academic Publishers, Boston, 111-151.

[20] Fischer, A, and Moncrief, V (1994b), *Classical and conformal superspace, linearization stability, and the reduction of Einstein's equations*, in *Proceedings of the Cornelius Lanczos International Centenary Conference*, J. Brown, M. Chu, D. Ellison, and R. Plemmons, editors, Society for Industrial and Applied Mathematics, Philadelphia, 535-542.

[21] Fischer, A, and Moncrief, V (1995a), *A method of reduction for Einstein's equations of evolution and a natural symplectic structure on the space of quantum gravitational degrees of freedom*, General Relativity and Gravitation.

[22] Fischer, A, and Moncrief, V (1995b), *Quantum conformal superspace*, General Relativity and Gravitation.

[23] Fischer, A, and Tromba, A (1984a), *On a purely "Riemannian" proof of the structure and dimension of the unramified moduli space of a compact Riemann surface*, Mathematische Annalen **267**, 311-345.

[24] Fischer, A, and Tromba, A (1984b), *Almost complex principle fiber bundles and the complex structure on Teichmüller space*, J. für die reine und angewandte Mathematik **352**, 151-160.

[25] Fischer, A, and Tromba, A (1984c), *On the Weil-Petersson metric on Teichmüller space*, Trans. Amer. Math. Soc. **284**, 319-335.

[26] Fischer, A, and Tromba, A (1987), *A new proof that Teichmüller space is a cell*, Trans. Amer. Math. Soc. **303**, 257-262.

168

[27] Freed, D, and Groisser, D (1989), *The basic geometry of the manifold of Riemannian metrics and of its quotient by the diffeomorphism group*, Michigan Math. J. **36**, 323-344.

[28] Friedman, J (1990), *Space-time topology and quantum gravity*, in *Conceptual problems in quantum gravity*, A. Ashtekar and J. Stachel, editors, Birkhäuser, Boston, 539-572.

[29] Friedman, J, and Witt, D (1986), *Homotopy is not isotopy for homeomorphisms of 3-manifolds*, Topology **25**, 35-44.

[30] Freedman, M, and Yau, S-T (1983), *Homotopically trivial symmetries of Haken manifolds are toral*, Topology **22**, 179-189.

[31] Gil-Medrano, O, and Michor, P (1991), *The Riemannian manifold of all Riemannian metrics*, Quart. J. Math. Oxford **(2) 42**, 183-202.

[32] Giulini, D (1994), *3-manifolds for relativists*, International Journal for Theoretical Physics, **33**, 913-930.

[33] Giulini, D (1995), *On the configuration-space topology in general relativity*, Helvetica Physica Acta, to appear.

[34] Gribov, V (1976), *Quantisation of non-Abelian gauge theories*, Nucl. Phys. **B139**, 1-19.

[35] Hatcher, A (1976), *Homeomorphisms of sufficiently large P^2-irreducible 3-manifolds*, Topology **15**, 343-347.

[36] Hawking, S, and Ellis, G (1973), *The Large Scale Structure of Space-Time*, Cambridge University Press, Cambridge, England.

[37] Hempel, J (1976), *3-manifolds*, Annals of Mathematics Studies, Number 86, Princeton University Press, Princeton, New Jersey.

[38] Hsiang, W (1967a), *The natural metric on $SO(n)/SO(n-1)$ is the most symmetric metric*, Bull. Amer. Math. Soc. **73**, 55-58.

[39] Hsiang, W (1967b), *On the bounds on the dimensions of the isometry groups of all possible riemannian metrics on an exotic sphere*, Ann. of Math. **85**, 351-358.

[40] Hsiang, W (1971), *On the degree of symmetry and the structure of highly symmetric manifolds*, Tamkang Journal of Mathematics, Tamkang College of Arts and Sciences, Taipei, **73**, 1-22.

[41] Hu, S-T (1959), *Homotopy Theory*, Academic Press, New York.

[42] Itoh, M (1991), *Yamabe structures and the space of conformal structures*, International Journal of Mathematics **2**, 659-671.

[43] Kneser, H (1929), *Geschlossen Flächen in dreidimensionalen Mannigfaltigkeiten*, Jber. Deutsch. Math.-Verein **38**, 248-260.

[44] Lelong-Ferrand, J (1969), *Transformations conformes et quasiconformes des variétés riemanniennes; application à la démonstration d'une conjecture de A. Lichnerowicz*, C. R. Acad. Sci. Paris **269**, 583-586.

[45] Lelong-Ferrand, J (1971), *Transformations conformes et quasiconformes des variétés riemanniennes compacts (démonstration de la conjecture de A. Lichnerowicz)*, Acad. Roy. Belg. Cl. Sci. Mem. Coll. 8°(2) **39**, no.5.

[46] Massey, W (1967), *Algebraic Topology: An Introduction*, Harcourt, Brace and World, New York.

[47] Mess, G (1995), *Homotopically trivial symmetries of 3-manifolds are toral*, to appear.

[48] Milnor, J (1956a), *Construction of universal bundles: I*, Ann. Math., (2)**63**, 272-284.

[49] Milnor, J (1956b), *Construction of universal bundles: II*, Ann. Math., (2)**63**, 430-436.

[50] Milnor, J (1961), *A unique decomposition theorem for 3-manifolds*, American Journal of Mathematics, 1-7.

[51] Mitter, P, and Viallet, C, (1981), *On the bundle of connections and the gauge orbit manifold in Yang-Mills theory*, Commun. Math. Phys. **79**, 455-472.

[52] Moncrief, V (1976), *Space-time symmetries and linearization stability of the Einstein equations. II*, J. Math. Phys. **17**, 1893-1902.

[53] Moncrief, V (1989), *Reduction of the Einstein equations in 2+1 dimensions to a Hamiltonian system over Teichmüller space*, J. Math. Phys. **30** (12), 2907-2914.

[54] Moncrief, V (1990), *How solvable is (2+1)-dimensional Einstein gravity?*, J. Math. Phys. **31** (12), 2978-2982.

[55] Munkres, J R (1960), *Obstructions to smoothing piecewise-differentiable homeomorphisms*, Ann. of Math. **72**, 521-554.

[56] Obata, M (1971), *The conjectures on conformal transformations of Riemannian manifolds*, J. Diff. Geom. **6**, 247-258.

[57] Omori, H (1970), *On the group of diffeomorphisms of a compact manifold*, Proc. Symp. Pure Math., Amer. Math. Soc. **15**, 167-183.

[58] Orlik, P (1972), *Seifert manifolds*, Lecture Notes in Mathematics **291**, Springer-Verlag, New York.

[59] Orlik, P, and Raymond, F (1968), *Actions of SO(2) on 3-manifolds*, in Proceedings of the Conference on Transformation Groups, New Orleans, 1967 Springer-Verlag, New York.

[60] Palais (1966), *Homotopy theory of infinite dimensional manifolds*, Topology, **5**, 1-16.

[61] Raymond, F (1968), *Classification of the action of the circle on 3-manifolds*, Trans. Amer. Math. Soc., **131**, 51-78.

[62] Satake, I (1956), *On a generalization of the notion of manifold*, Proc. Nat. Acad. Sci. U.S.A. **42**, 359-363.

[63] Scott, P (1983), *The geometries of 3-manifolds*, Bull. London Math. Soc., **15**, 401-487.

[64] Singer, I (1978), *Some remarks on the Gribov ambiguity*, Commun. Math. Phys. **60**, 7-12.

[65] Sorkin, R (1989), *Classical topology and quantum phases: quantum geons*, in *Geometrical and Algebraic Aspects of Nonlinear Field Theory*, S. De Filippo, M. Marinaro, G. Marmo, and G. Vilasi, editors, Elsevier Science Publishers, 201-218.

[66] Spanier, E (1966), *Algebraic Topology*, McGraw-Hill Book Company, New York.

[67] Thurston, W (1978), *The geometry and topology of 3-manifolds*, preprint, Princeton University, Princeton, New Jersey.

[68] Thurston, W (1982), *Hyperbolic geometry and 3-manifolds*, in *Low-dimensional topology*, London Mathematical Society Lecture Note Series 48, R. Brown and T. L. Thickstun, editors, Cambridge University Press, 9-25.

[69] Thurston, W (1982), *Three dimensional manifolds, Kleinian groups and hyperbolic geometry*, Bull. Amer. Math. Soc. **6**, 357-381.

[70] Tromba, A (1992), *Teichmüller Theory in Riemannian Geometry*, Birkhäuser Verlag, Basel.

[71] Waldhausen, F (1967), *Gruppen mit Zentrum und 3-dimensionale Mannigfaltigkeiten*, Topology **6**, 505-517.

[72] Waldhausen, F (1968), *On irreducible 3-manifolds which are sufficiently large*, Ann. Math. **18** 56-88.

[73] Wheeler, J A (1962), *Geometrodynamics*, Academic Press, New York.

[74] Wheeler, J A (1964), *Geometrodynamics and the issue of the final state*, in *Relativity, Groups and Topology*, C. DeWitt and B. DeWitt, editors, Gordon and Breach Science Publishers, New York.

[75] Wheeler, J A (1968a), *Superspace and the nature of quantum geometrodynamics*, in *Batelle Rencontres - 1967 Lectures in Mathematics and Physics*, C. DeWitt and J. A. Wheeler, editors, W. A. Benjamin, Inc., New York.

[76] Wheeler, J A (1968b), *Einstein's Vision*, Springer-Verlag, New York.

[77] Wheeler, J A (1970), *Superspace*, in *Analytic Methods in Mathematical Physics*, R. Gilbert and R. Newton, editors, Gordon and Breach Science Publishers, New York.

[78] Whitehead, J H C (1961), *Manifolds with transverse fields in euclidean space*, Ann. of Math. **73**, 154-212.

[79] Witt, D (1986), *Symmetry groups of state vectors in quantum gravity*, J. Math. Phys. **27** (2), 573-592.

[80] Wolf, J A (1972), *Spaces of Constant Curvature*, second edition, Publish or Perish Press, Berkeley, California.